Studies in Systems, Decision and Control

Volume 5

Series editor

Janusz Kacprzyk, Polish Academy of Sciences, Warsaw, Poland
e-mail: kacprzyk@ibspan.waw.pl

For further volumes:
http://www.springer.com/series/13304

About this Series

The series "Studies in Systems, Decision and Control" (SSDC) covers both new developments and advances, as well as the state of the art, in the various areas of broadly perceived systems, decision making and control- quickly, up to date and with a high quality. The intent is to cover the theory, applications, and perspectives on the state of the art and future developments relevant to systems, decision making, control, complex processes and related areas, as embedded in the fields of engineering, computer science, physics, economics, social and life sciences, as well as the paradigms and methodologies behind them. The series contains monographs, textbooks, lecture notes and edited volumes in systems, decision making and control spanning the areas of Cyber-Physical Systems, Autonomous Systems, Sensor Networks, Control Systems, Energy Systems, Automotive Systems, Biological Systems, Vehicular Networking and Connected Vehicles, Aerospace Systems, Automation, Manufacturing, Smart Grids, Nonlinear Systems, Power Systems, Robotics, Social Systems, Economic Systems and other. Of particular value to both the contributors and the readership are the short publication timeframe and the world-wide distribution and exposure which enable both a wide and rapid dissemination of research output.

Kofi Kissi Dompere

Fuzziness, Democracy, Control and Collective Decision-Choice System: A Theory on Political Economy of Rent-Seeking and Profit-Harvesting

 Springer

Kofi Kissi Dompere
Department of Economics
Howard University
Washington District of Columbia
USA

ISSN 2198-4182   ISSN 2198-4190   (electronic)
ISBN 978-3-319-38286-9   ISBN 978-3-319-05329-5   (eBook)
DOI 10.1007/978-3-319-05329-5
Springer Cham Heidelberg New York Dordrecht London

Printed on acid-free paper

Springer is part of Springer Science+Business Media (www.springer.com)

Dedication

To the global children whose future Rests firmly in the hands of social thinkers
And practitioners of thinking system of just order for peace;

To Mother Earth whose past, present and future glory
Lies in the uniting forces of your children in mind, body and sprit and
In triumph of peace over war; Justice over injustice and collectivity over
individualism;
To the global poor whose struggle against Imperial oppression, neocolonial
injustice and racism
Is a constant reminder of never-ending search for global golden tomorrow
through the strength of Human spirit.

To all scholars who are hard at work for the construction of thought system
For justice and reconciliation without which Peace has no anchorage; and for
the Understanding of political deception and destruction of sovereignty
That is the core of true democracy and the people's participation.
To all the media persons who are working hard to expose corruption of the
practices of the members of the governing to enrich themselves; and to all persons
working hard
In order to strengthening democracy and rescue it from the mirage of the desert;

As a strategy of full spectrum resistance against a full spectrum dominance
Of bureaucratic capitalism and bureaucratic imperialism that seek
The destruction of human strength, courage, hope and wisdom;
AND
To all those seeking to emancipate themselves from cognitive imbecility,
And the zone of the familiar with the practice of methodological doubt and
hope.

Acknowledgements

I wish to express my thanks to all my friends, who one way or the other, have given me encouragement and emotional support as I tried to clear some logical haze surrounding domestic and global political economy and foundations of economic theory. My special thanks also go to my graduate students who accepted the challenges of the development of critical thinking by developing the power to question the accepted in other to avoid being shut out from emerging possibilities as well as being victimized by the intellectual walls of the familiar. I hope I have been able to instill the principle of cognitive audacity, curiosity and creativity in at least some of them. I give my gratitude to Professor Kwabena Osafo-gyimah for his continual encouragement. I give thanks to the Hall Masters of Akuafo Hall at the University of Ghana and the staff for their generosity in hosting me during my visits to Accra to write and tighten some ends of the development of the theories on political economy. I also express great thanks to all authors who have been referenced and to those authors who have enriched my philosophical, mathematical and analytical skills. Thanks to Ms. Jasmine Blackman for her editorial suggestions.

Controversial ideas and terminologies are intentional, and intentionally directed to restructure the paradigm of contemporary theories on political economy under collective democratic decision-choice system where defective information structure, composed of vagueness and incompleteness, and deceptive information structures, composed of disinformation and misinformation, are the characteristic inputs of individual and collective thinking that has become problematic in democratic decision-choice space whose outcomes invariably define the path of peoples' progress and history.

I hope the introduction of deceptive information structure into optimal collective decision-choice activities in the political economy will be useful in explaining decision-choice outcomes in political economies and the cultivation of secrecy and the use of principles of fear and public safety by the governing class to dismantle popular discontent. I also hope that the use of fuzzy paradigm, its logic, laws of thought and mathematics will offer an approach to the use of qualitative mathematics in theory construction and reduction in socio-political transformations including economic development. I accept responsibility for any error that may arise in the arguments.

Preface

And I came before the Master and I said, "What can I do to be saved?" And he said to me "SEEK YE FIRST THE POLITICAL KINGDOM AND ALL OTHERS SHALL BE ADDED UNTO IT." And I said to the Master, "Jesus Christ said 'Seek ye FIRST the kingdom of God, and all others shall be added to it.' Which came FIRST?" And the Master said to me, "Jesus also said 'Render unto Caesar the things that are Caesar's and unto God the things that are God's.'"
And wisdom of the Master unfolded before me...

Kofi Ghanaba (Guy Warren of Ghana) in *I Have Story to Tell*, Accra, Ghana 1962, p.2

In the advanced capitalist nations, new elites based on science and technology are gradually displacing the older elites based on wealth.

(Robert L. Heilbroner : The New York Times Magazine)

The notion that power in the modern economy lies increasingly with the great organizations and increasingly less with the supposedly sovereign consumer and citizen has also been making its way into the text books. Something here is owing to a vacuum. In recent years there has been a rapidly growing discontent with the established or neoclassical model of economic and political life. The way was open for an alternative. Still the inertial forces are great. The textbook writer is naturally a cautious fellow. Like liberal candidates for public office he must always have one eye for what is reputable and salable as distinct from what is true. John Kenneth Galbraith [R11.23 p. xii].
Some of these organizations are very large; as few would doubt, they have power, which is to say they can command the efforts of individuals and the state. John Kenneth Galbraith [R11.23, p.3]

The rationally casually ignorant voter is a very slender reed on which to build the foundations of democratic politics. He is much more likely to be the recipient of the dispersed costs than of the concentrated benefits of the legislative process. He is much more likely to suffer the net costs of random prisoners' dilemmas than to enjoy the systematic gains-from-trade outlined in the "Calculus of Consent". Is it legitimate, in such circumstances to infer that the forces of supply and demand in political markets are driven not by individual voters but by interest groups; that

*collective action replaces individual action in the battle over the spoils of politics
that is the raison d'être of democratic politics? If so, what predictions can be
made about the rent-seeking consequences of competition among pressure groups
for political influence?* Gordon Tullock, [R13.6, p.46]

*The fundamental category of economic activity is power. It is equally (because
in both cases completely) essential to any good activity to have power and to
direct its use "rightly" – and in particular not to treat the acquisition of power as
an end or purpose on its own account.*

Frank H. Knight (On the History and Method of Economics, p. 30.)

*Great and irresponsible power is a threat to any civil society, and the processes
by means of which this power is gained and exercised tend to corrupt the
democratic institutions of government.* B.S. Keirstead, [R15.21, p.445]

I: The Monograph

This is a second volume in my treatment on the problems in a political economy
where the social decision-choice actions are framed in terms of a democratic
collective decision-choice system and where the political economy is seen to stand
on the three legs of economic, political and legal structures. The first volume was
devoted to examining the problems of the political economy of social goal-
objective formation under democratically majoritarian principles, and how the
problems are related to the relative private-public sector provision of goods and
services in the society through the goal-objective formation. The central focus
accorded to the social goal-objective formation is due to an analytical position that
the contemporary social systems are organized to accomplish national interest and
implicit or explicit social vision which are supported by the outcomes of the goal-
objective formation. The relative public-private sector provision of goods and
services and the nature of social income distribution are instruments for
administering the elements in the social goal-objective set under market or non-
market institutions.

The problem of the relative public-private sector provision of goods and
services and the nature of social income distribution is an integral part of the
theory of organizational efficiency and justice in cost-benefit distribution. In
traditional economic theory and analysis, the social vision, national interest and
social goal-objective set are implicitly assumed on the basis of which the quality
of the social organism and social welfare are determined. Given this foundational
assumption, the economic decision problem is simply what are the best allocation
of the resource endowment, best production of goods and services, and the best
distribution of costs and benefits of goods and services to the people on the
fundamental principles of a democratic collective decision-choice system. The
political decision problem is what should be the best institutional configuration
that will support the solution to the economic problem.

The current volume is about the analysis of the problems and solutions of the
market mockery of the democratic collective decision-choice system under a

public information constraint where the voters dwell in the sphere of phantom power. The market mockery is related to the nature of the power distribution of productive factors in the political, economic and legal structures which form the foundation of the political economy operating under a market mechanism. The market mockery is seen in terms of the formation of the social goal-objective set and how the elements are related to private-public sector provision of goods and services and its connections to private-public-sector conflicts which generate conditions of rent-seeking in the political structure, rent protection in the legal structure and rent harvesting for profit enhancement in the economic structure. The monograph is essentially about the political economy of rent-seeking, rent-protection and rent-harvesting to enhance profits under a democratic collective decision-choice system, and the market mechanism where the relative private-public sector production of goods and services, income distribution and wealth distribution through prices, taxes and consumption of goods and services are the instruments of social policies in the political, legal and economic structures for income transfers.

The political structure, the economic structure and the legal structure with their corresponding markets and powers are analytically distinguished and examined in relation to the democratic collective decision-choice system of the political economy. The roles that individual and group resources, money and information play in the decision-choice outcomes are examined analyzed and related to the political economy of rent-seeking which is then formulated as a game played between powerful private concerns and the government. The rent-seeking game is partitioned into a rent-creating game with a creating strategy, rent-protection game with a protection strategy and a rent-harvesting game with a rent-harvesting strategy. The environment of these gaming processes is defined by an imperfect information structure composed of defective and deceptive information sub-structures leading to a systemic risk that is composed of fuzzy and stochastic risks of an aggregative behavior. Two classes of rent-seekers are identified. They are rent-seeking innovators with rent-creating investments to create rent-seeking environments of opportunities in the political and legal structures, and rent-harvesters with rent-extracting investments that abstract rent in the economic structure and deplete the rent opportunities in the political and legal structures. The continual interaction between the two classes maintains the dynamics of the rent-seeking process. The rent-seeking game continually changes or endorses the governors that use the state power. The umpire of the rent-seeking games is the national constitution which is also under the control of the political structure. These games are specified and analyzed around the elements in the social goal-objective formation, national interest and social vision. The collective decision is related to the problems of establishing the appropriate government size and the required relative private-public sector combination in the provision of goods and services in the political economy.

The nature of the rent-seeking games and activities are amplified by examining them in terms of Schumpeterian and Marxian socio-political dynamics along the private-public-sector efficiency frontier that is connected to continual conflicts of freedoms in individual-collective duality. The properties of Schumpeterian and

XII   Preface

Marxian political economies are specified to highlight their similarities and differences in relation to rent-seeking games that are related to the income-distribution game. The similarities and differences are related to the power distribution and decision-choice actions on private-public sector provision of goods and services which are illustrated with organizational polar cases for the understanding of the process of rent-seeking, rent-preservation, rent-harvesting, profit enhancement and ideological conflicts.

The rent-seeking activities are divided into rent-seeking in the real sector composed of labor and commodity markets, and rent-seeking in the financial sector composed of money and debt markets. The activities of the two sectors are related to conditions of wealth-creating rent-seeking on one hand and non-wealth-creating rent-seeking on the other hand. The financial-sector goods and services constitute *debt* instruments that have claims over real commodities and services. The real-sector goods and services constitute the *collateral* for the financial sector, it is argued. The non-wealth-creating rent-seeking activities are argued to relate to income-wealth transfers that give rise to financial bubbles through fractional banking, and in the sense that the financial bubbles have no collateral support in the real sector. The financial bubbles increase the systems risk depending on the disparity between the values of the real-sector production and the financial-sector production. The disparity creates sectorial disequilibrium and Schumpeterian transformational dynamics in the general markets of the political economy. The conditions of the stability of the transformational dynamics are examined in relation to the class political power struggles between the owing classes of capital and labor. The real and financial sectors are connected by the principle of collateralized actions through the cost of borrowing which relates the rent-seeking activities in the two sectors.

The rent-seeking activities in the two sectors are related to the conditions of the democratic collective decision-choice system and examined in relation to the political market, responsibility, accountability and governmental budgetary decisions, and then to the voting processes in the social decision-choice set-up. The budgetary process is viewed as a game process with or without deceptive information structure and then related to the implementation of decision-choice actions of the elements in the goal-objective set where the budgetary-voting games are shaped by the private-public sector ideological positions. The conditions of corruption are examined under political duopoly in terms of accountability-corruption games. The monograph is concluded by examining the interactions among rent-seeking activities, penumbral regions of decision-choice actions, fuzzy-decision-choice rationality and the nature of constitutional dilemmas in democratic social formations where the market generates conditions of mockery of democracy and the maintenance of phantom power for the masses.

II: The Objective of the Monograph

The monograph is thus about the political economy of rent-seeking in democratic social systems where social decision-choice actions are guarded by the majoritarian principle of vote casting under relevant information constraint. The concept of rent-seeking is explicated in terms of manipulation of the institutions of government and the members who administer them with powerful groups in order to carve out special privileges for themselves under some created conditions in the political and legal markets. The net cost-benefit balance of the created privileges is negative to the society, where the costs of created privileges are socialized to the least powerful and the benefits of the privileges are privatized to the most powerful within the political economy under a capitalistic market mechanism. The process is simply the socialization of costs of private privileges and privatization of benefits of social privileges of the production-distribution activities in the institutional trinity of economics, politics and law. The primary purpose of this monograph is to facilitate the understanding of this process and the explanation of the differential outcomes of this rent-seeking process and show how the understanding and explanation will constitute the foundations for explaining the private-public-sector conflicts and social policy in the democratic capitalist social formation for production, distribution and consumption wrapped in the capsule of the fundamental ethical postulate of individual freedom as opposed to collective freedom. There are other supporting objectives of this monograph.

1. To explain the forces behind the movement along the relative private-public sector efficiency frontier, defined in terms of provision of social goods and services, and then examine the strengths and weaknesses of the privatization debate and the role of economic, political and legal markets in the mockery of democracy and sovereignty.

2. To examine the claim that government no longer has a valid claim to determine the values, national interests, national vision and a set of priorities through the formation of the social goal-objective set for society because the market system can and should undertake these social activities.

3. To examine the claim and criticism that the capitalist's democracy is a political economy where the economic power, political power and legal power are effectively controlled by the capital-owning class in such a way as to control the state power which is then used by the same capital-owing class as the police arm at its disposal to directly or indirectly regulate behaviors in the economic, legal and political structures in support of its class interests, rent-seeking activities for profit enhancement and distortion of income distribution.

4. To examine the role of ideology and deceptive information structure in shaping activities of rent-seeking in transformational dynamics in the Schumpeterian and Marxian political economies in relation to regulation, governance, accountability, socio-economic policies and the governmental budgetary process.

5. To examine the conditions under which the interests of powerful business concerns become isomorphic to the national interests and social vision, and where

the state power is activated the powerful concerns to use the national military power in support of their interests, since the interests of the powerful concerns are the same as the national interests and social vision. Furthermore, to examine the conditions under which the powerful business concerns directly or indirectly use the state power to reduce the labor power.

III: An Approach to the Study of the Political Economy in the Monograph

The toolbox used in the monograph consists of methods of fuzzy decision, approximate reasoning, negotiation games and fuzzy mathematics. The use of this approach to the understanding of the political economy is made possible by first rejecting the neo-classical income-distribution paradigm based on the technical conditions of production where factors claim their contributions at the margin of production. The marginal productivity theory of income distribution does not help us to explain an increasing disparity in income distribution that may be attributed to institutional inefficiencies, changes and distortions due to the presence of defective information structure, subjective phenomena and power asymmetry in the application of rules and regulations in the political and legal structures. It also rejects the postulate of perfect and exact information structure. In place of the postulate of information perfection and exactness, we have the postulate of defective and deceptive information structures in the decision-choice system.

The introduction of deceptive information structure is relevant for understanding perceptions of distrust and commitments of opposing powers. It further points to the notion that free to choose in the theoretical structure elegantly built by the neo-classical economists for the consumer decision-choice system, the producer decision-choice system, the general equilibrium system and welfare economics, have an underlying assumed institutional and power systems whose instabilities, due to power shifts, change the qualitative disposition of the choice variables to affect the individual and collective preferences for any form of optimal behaviors. In this respect and at the level of the theory of the political economy, the task of the logical construct is to search for a general principle of administration and development of the institutional trinity of economics, politics and law and its impact on society and the social life of production and reproduction of real life under conditions of unstable static and temporary equilibrium, which will define the dynamic equilibrium-disequilibrium process and path of the political economy.

IV: The Organization of the Monograph

The structure of the discussions and the theoretical framework requires general and specific tools that will be useful in dealing with the collective decision-choice system under democracy, power systems, and defective and deceptive information structures. The analytical tools used here include methods of fuzzy decision theory, negotiation games to formulate and analyze the problems and solution sets of social opposites with relational continuum. Chapter one presents the essential structure of the hermeneutics of rent-seeking theory and how it relates to social goal-objective formation. Chapter two poses the problems and structure of the rent-seeking system including rent-creation, rent-protection and rent-harvesting in relation to profit enhancement and social goal-objective formation that define the analytical structure. Chapter three discusses the rent-seeking activities in the Schumpeterian and Marxian political economies and their relationships to the capital-labor power struggle for institutional transformation and the control of the government. Chapter four is used to examine the relational nature of the financial bubble, systemic risk and creative destruction as part of the rent-seeking behavior in the political economy, and the dynamic role that rent seeking plays in the Schumpeterian political economy.

In Chapter five, critical examinations are undertaken to discuss the relational interplay of democracy, political responsibility, accountability in relation to interests of individuals, constituency, party affiliation and nation under the conditions of the existence of a non-market and market in the institutional trinity, and its relationship to the power trinity. The discussions are further extended to examine the relational structure of rent-seeking and public-private sector dynamics within the framework of national interest, social vision and the goal-objective formation. The monograph is concluded with Chapter six which is used to examine the penumbral regions of the rent-seeking decisions given the political markets in the presence of defective and deceptive information structures.

The analytical work that is being undertaken from chapter one to chapter six in this monograph may be viewed as founded on a number of conceptual building blocks. These building blocks may be viewed separately for analytical convenience but they are relationally connected to assert the outcome of the elements in the system in a general disequilibrium-equilibrium dynamics. We have a notion of institutional trinity composed of economic, political and legal structures. The institutional trinity has interdependency relation with a power trinity composed of state, labor and capital powers. There is a set of decision-choice agents, composed of either the individuals or the collective with criteria of behavior which are made up of self-interests defined in the cost-benefit space under either market or non-market conditions, where the individual and the collective struggle for the best with defined power constraints on the sovereignty. Here, each decision-choice agent is viewed as composed of a trinity of homo *economicus*, homo *politicus* and homo *legalicus*. The same decision-choice agent is the economic being, political being, and legal being where the behavior in the political structure changes the qualitative dynamics of the legal structure to affect both qualitative and quantitative decision-choice variables in the economic structure. Thus a general decision-choice agent who is three in one must coordinate the three decision-choice elements, the associated risks and possible outcomes, and the cost-benefit implications in any decision-action. The decision-choice agent must consider also the relational implications on the other structures of a decision-choice action in any of the three structures. It is the decision-choice behavior of this relational unity that a complete economic theory must deal with. It is also this relational unity that shapes the outcomes and cost-benefit implications of democratic collective decision-choice systems. Moreover, the distribution of the unified cost-benefit implications in the collective democratic decision-choice systems also brings about a power struggle and the manufacturing of deceptive information structure within the justice-injustice duality. The relational unity of all the possible decisions is presented in terms of a cognitive geometry in Figure IV(1).

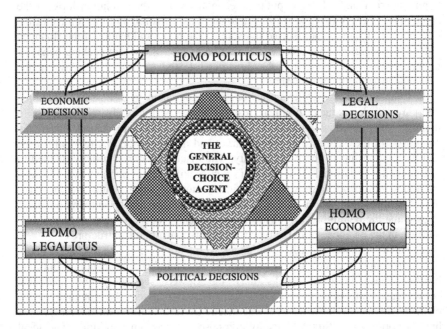

Fig. IV (1)1: Cognitive Geometry of Relational Structures in Defining the Nature of the General Decision-Choice Agent in the Political Economy with Unity of Action in the General Decision-Choice Space

The decision-choice agent must balance costs and benefits in the economic, political and legal structures under the available information structure. In this frame, the political economy generates two kinds of information structures that produce total uncertainty. The uncertainties are composed of qualitative and quantitative information restrictions to produce knowledge with the conditionality of fuzziness and incompleteness. Also in this frame, the classical optimization methods fail to provide us with reasonable understanding of the individual risk and the systemic risks associated with the fuzzy-stochastic behavior generated by the aggregate outcomes of the interactions of individuals in the collective decision-choice space with power and information asymmetry. However, the use of fuzzy paradigm, composed of its logic and mathematics, provides us with analytical methods to frame the decision-choice problems under such an information constraint and abstract solution.

Existing economic theory on optimal decision is devoted to the study of a representative decision-choice agent as a homo *economicus* who has nothing to do with the political and legal structures. Similarly, the political theory concentrates on a decision-choice agent as homo *politicus* and neglects the aspects related to legal and economic decisions. In a similar fashion and at the level of decision-making, legal theory does not examine the decision-choice agent and his or her relationship with the economic and political decisions. Instead, much of the legal theory is seen in terms of the philosophy of law that is devoted to the study of the nature and structure of laws in relation to liberty, rules, justice, fairness, power, order, responsibility accountability, punishment and obedience that constitute the defining structure of the environment in which decision-choice agents operate in the political economy. The decision-choice agent as a homo *legalicus* whose choices are directed to cost-benefit configuration in the economic structure is dismantled from the other structures. The analytical work is to unify the three components of the same decision-choice agent in a fuzzy-constrained collective decision-choice process under the principle of democracy. Here, it is useful to pay attention to Professor Joan Robinson's statements:

In general mass of notions and sentiments that make up an ideology those concerned with economic life play a large part, and economics itself (that is the subject as it is taught in universities and evening classes and pronounced upon in leading articles) has always been partly a vehicle for the ruling ideology of each period as well as partly a method of scientific investigation [R15.41 P.1]... Since the egoistic impulses are stronger than the altruistic, the claims of others have to be imposed upon us. The mechanism by which they are imposed is the moral sense or conscience of the individual [R15.41, p.5]... Any economic system requires a set of rules, an ideology to justify them, and a conscience in the individual which makes him strive to carry them out [R15.41, p.13]... The lack of an agreed and accepted method for eliminating errors introduces a personal element into economic controversies which is another hazard on top of all the rest. [R15.41, p. 25].

The statements by Professor help to show the essential work in this monograph which I hope will be useful addition to the literature on rent-seeking, national goal-objective formation, national-interest definition, social-vision determination and institutional development as important foundations in understanding resource allocation, output production and income distribution in any domestic political economy and how it relates to the international political economy.

Prologue

I: Decision and Choice in Political Economy

Every aspect of human action is decision-choice based starting from conception to implementation under any given information regime and institutional arrangement. The relationships among the decision agent, information, power, decision and choice constitute the decision-choice system which includes the social sep-up. The decision-choice system is simpler when it involves personal actions. The simplicity emerges from the condition that the decision-making power is vested in the beneficiary who also bears the costs of the decision-choice action. In this case, the principal and the agent are the same in one and hence we have a unified set of preferences in all structures of decision-choice actions in relation to any social cost-benefit configuration. Complications tend to arise when the decision-choice system involves more than one individual in which case we have a multi-person situation and hence confronted with the problem of reconciling conflicting preferences to arrive at a collective preference in a cost-benefit space except when all the individuals have the same preferences. This multi-person problem is further complicated when we must deal with multiobjective and multi-criteria situations where the decision-choice system is self-exciting and self-correcting. The decision-choice system is further complicated when the information regime is defined by *defective* and *deceptive* information structures. These problem conditions call for the use of generalized theory of decision-choice system that must deal with the defective and deceptive information structures. Such a decision-choice system involves, among other things, the policy events in the political economy and their implementations at the organizational presence of government, power, sovereignty, ideology and many important relational actions.

II: The Government, The Governance and Their Relational Structure

All political economies relate to resources, production, goods services and decision-making systems. In other words, the political economy is about the decision-choice management of cost-benefit configurations in the social space in time and over time. The social decision-making system involves power relations that are directed to resource management, resource allocation and production of goods and services, and the distribution of these goods and services. The allocation of the decision-making power and the possible relations among the members is a problem of all societies. It is the solution to this power allocation problem that induces the rise of governments and the construct of various forms of political economy under collective existence to reconcile conflict in the individual-community duality. The government is thus a facility available to the social decision-making core. This facility is a created institution in which a social power is vested by the sovereign people to make social decisions that affect all the members of the collective. The social decision-choice actions regarding national interest and social vision must reflect the preferences of the people as decided or as much as they can be determined. The government as a social instrument is endowed with social power to craft rules, regulations, laws and instruments to enforce compliance in order to resolve conflicts in preferences and create social stability in the process of resource allocation, production and fairness of distribution. Government is thus viewed as part of social institutional arrangement in which power is vested to mediate conflicts in the distribution and the uses of power. Corresponding to the government is the *social decision-making core* which is charged with a mandate to use the government to govern. While the government is of the people, by the people and for the people, the governance is not necessarily of the people, by the people and for the people. The social decision-making core may be constructed in any manner a particular social system may select. In this respect, the study of the political economy is the study of power distribution in relation to any collective decision-making system of politics, law and economics. There is a clear distinction between the concepts of government and governance.

The government is thus institutional relations of people, sovereignty, autonomy and power systems of decision making. With the political economy, sovereignty may be seen in terms of decision-making authority with various guiding principles as to how, who, where and when such an authority is conferred and exercised. The decision-making authority relates to the autonomy to act and to the acceptance of responsibility of the outcome. In the political economy, the problems of individual and collective decision-choice autonomies are in reference to the applications of sovereignty and power over the institutional trinity of economic, legal and political structures. The relationship of the institutional trinity to the individual and the collective in terms of preferences tends to establish the nature of the struggles in the individual-community priorities and hence the integrity of individual-self or the collective-self in relation to the distribution and the potential uses of the decision-making power that one possesses in the political economy. Here, it may be kept in mind that freedom implies power of decision-making in the political economy. However, power without restraint leads to abuse and unchecked restraint leads to a lost of freedom and sovereignty. The social decision-choice problem is to seek and create a balance of institutional configuration within the private-public sector duality for any given point of time.

At the level of application and uses of the decision-making power, the government is charged with the responsibility of constructing national policy that is related to national interests, social vision and the supporting set of social goals and objectives under the constraint of the citizens' sovereignty. It is the definition of national policy in relation to national interest and social vision with the creation of the supporting social goals and objective set that conflicts arise in the individual and collective preferences that require some technique of computational aggregation over the collective decision-choice space. It is also here that intense political games arise among groups that may take the form of compromises in favor of private sector and against the public sector with varying degrees of political corruptions for rent-seeking and profit or benefit enhancements. A similar situation arises when the public sector is viewed to represent collective interest. The operational situation is that all national policy constructs require balancing conflicts in individual preferences in the collective decision-choice space leading to needed compromises that can support the general social interest rather than compromises between powerful interest groups for their special and private interests at the expense of the collective interest and national good. In this respect, democracy in the collective decision-choice space of the political economy loses its meaning of freedom and assumes the form of its opposite which is lack of freedom.

III: Social Power and the Government

It has been suggested in this monograph that the study of the political economy may be viewed in terms of the study of distribution of social power and the conflicts of the forces of power of the social decision-choice system in the three structures. The behavior of the decision-choice system is due to the nature of the distribution and the manner of use of the social decision-making power in the sectors of the institutional trinity and how the power distribution and utilization affect each sector and the possible sectorial relationships and interactions in terms of individual and collective cost-benefit balances in the political economy. The social decision-choice power takes the form of the trinity of *state power, labor power* and *capital power*. These powers have relative differential strengths in the economic, political and legal structures depending on the nature of the information structure, the general awareness of the sovereign individuals and the information-processing capacity of the power holders. In this way, the decision-choice system is defined by the interactive processes of the *power trinity* and *institutional trinity* where the center constitutes the conflict zone of the political economy with conflicting individual preferences for continual social transformations. The relational structure may be defined by two diagrams that point to the direction of the monograph, the task that is being undertaking and the analytical toolbox that is being employed. The diagram (III.1) shows the relational structure of the power system and organizational system. The political and legal structures are artificial creations to complement the economic structure for organization integrity of the social set-up. The economic structure is a natural creation in which every sovereign individual must participate for survival. However, the degree of freedom and the manner of participation in the economic structure are institutionally constrained from the political and legal structures that hold the collective constraining power. The people's power is the primary category from which labor power is a natural derivative. The labor power constitutes the organizational primary category from which the economic power is derived to institutionally give rise to the political power as a derivative and then serves as the primary category for the derivative of the legal power. This power system and institutional structures constitute the framework for the allocation of resources, production of goods and services and the distribution of social goods and services that are central to the classical and neoclassical microeconomic decision theory.

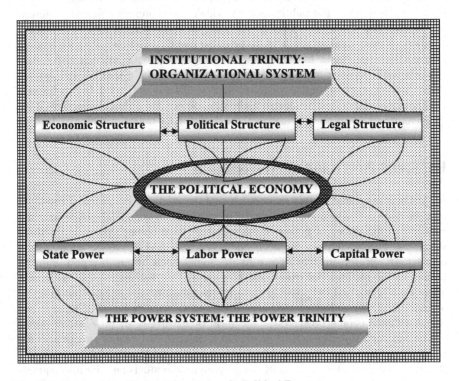

Fig. III.1 Power and Institutional Structures of a Political Economy

In support of this relational structure is the conflict zone of power-distributional uses which is illustrated in Figure III.2. It defines the environment of power games of negotiations for coalition formations to directly or indirectly control the institutions of government through their representatives. The cognitive geometry represents the form of the political economy as a decision-making system and the conflict zone of never ending struggle leading to the framework of politico-economic games and power struggles to control the political economy in terms of resource allocation, production of goods and services and distribution of the results of production in the time domain. The struggle to control the government is a struggle to control the economic structure in terms of ownership, resource allocation, production and distribution of goods and services among the members.

It is, thus, a struggle to control the social cost-benefit distribution of the social production. In the classical and neoclassical macro-micro economic decision-theoretic system the nature of the power distribution is implicitly assumed and the analyses concentrate on its uses. The economics of power distribution is not undertaken.

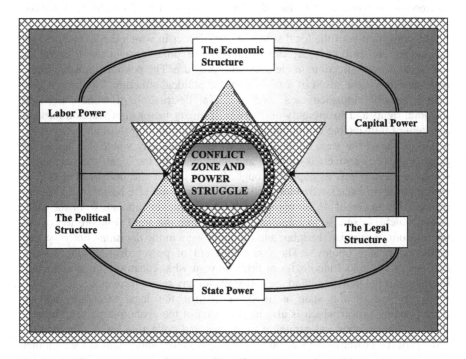

Fig. III.2 Decision-Making Power System with a Conflict Zone

IV: Sovereignty and the State Power

The power distribution is related to sovereignty where the state power is an institutional creation and is housed in the political structure and vested in the institutional facility of the government. The power of the state is a consolidation of power transfers from the sovereignty of individual decision-making powers. The power of the state in the political economy is created by the *principle of sovereignty transfers* where such transfers may be voluntary or involuntary. The labor and capital powers are defined by the principle of ownership of factors of production where such ownership may or may not be justified by the legal system. The ownerships of the factors of production bestow on the owners the decision-making power in terms of their uses. The labor and capital powers are defined in the economic structure and where there is a continual struggle between them for the control of the economic power. The analytical structure is such that the primary category of factors of production is labor where the ownership of labor is bestowed by natural creation and hence related to the fundamentally natural freedom to use by exercising the natural ownership right. Capital is a derived category of factor of production from labor. Its ownership is institutionally created and hence retains the derived freedom to use. In other words, the freedom to exercise ownership of capital is a derived freedom from the natural freedom of labor. The legal system is an instrument for mediation and hence it is a constraint on freedom to exercise either the labor or capital power by the individual and the collective.

The construction of the legal structure, the manner in which the laws, rules and regulations are enforced may create injustice or unfairness in the exercise of individual primary freedom in the economic structure. The possible injustice in the legal structure is the result of actions from the political structure. The presence of such a possible injustice is a seed of social destruction of some individual freedom; rights and sovereignty of use of the primary factor of production through a system of transferring of the individual rights of ownership to someone by force or the power of the state as in the case of slave economy where the individual primary factor is converted by force and maintained by the legal structure into a secondary factor, just as capital through the transfer of the individual decision-choice power to exercise freedom and rights to someone else besides the original owner. The power of decision-making by ownership is separate from the state power which is exercise to create the conditions of the legal structure to control the freedom to exercise the labor and capital powers in the decision-choice system in the political economy.   The control system of power distribution and use depends on the social ideologies as reflections of what constitutes a good society in relation to the individual and the collective for any social formation.

The power of the state is used to control the legal structure and its corresponding power which is also used to control the economic power which is shared between labor and capital at individual and collective levels. The control and use of the power of the state always reside in actual and potential vacuum that must be occupied either by the representation of labor class or its advocate or by the representation of capital class or its advocates or both in some proportion. The exercise of the state power to control the labor and capital powers in the economic structure is such that there is a continual struggle between the capital and labor owning classes and their advocates to control the state power through a process to occupy the actual or potential vacuum of the governmental vehicle. The legal structure and its institutions of control and enforcement create a transformation process in dualities involving freedom and justice, law and national security and transparency. Here, freedom without the constraint of law produces anarchy; and law without the balanced judicial practice of justice produces oppression. All these social dualities may be capture in a well-balance economic theory of power.

Disciplined individual and collective freedoms in the exercise of the decision-making powers require some form of controls from the legal structure and the controls require some lost of freedom. These controls are instituted to accomplish the effective implementations of the national interest, social vision and the elements of the social goal-objective set. In this case, the uses of sovereignty function in the continuum of duality of complete freedom and complete lack of freedom which may be mapped into a cost-benefit duality in the economic space.

The creation of institution of government and the voluntary or non-voluntary power transfers through the individual and collective sovereignties automatically establishes the problems of *principal-agent* duality that bring into the focus the *freedom-decision paradox* and *sovereignty-power paradox* which are the main problems of democratic social formation. Here, all individual sovereignties must count in the collective decision-choice system of the political economy. Ironically, the establishment of the principal-agent duality can lead to the affirmation of the cherished individual sovereignty and decision-making power. It can also lead to the creation of phantom individual sovereignty where the individuals are ideologically manipulated to believe that they have decision-making power that does not exist in their collective decision space. The conditions in which phantom power thrives is the electoral process for establishing the decision-making core as the resident of the government and as the agent for governance. In the essential social power structure, we may have electoral dictatorship and non-electoral dictatorship where in both cases the preferences of the citizens are subsumed under the preferences of the agent controlling the government.

V: Some Questions for Democratic Capitalism

In all the interactions of the power system and organizational system for the collective decision-choice system in the political economy, there are a number of questions that must be answered by democratic capitalism. What is the appropriate balance between the private and public sector's provision of goods and services in the society and does the problem of this appropriate balance have a permanent and stable solution? What are the best appropriate rules, regulations and laws to achieve a balance between individual freedom and collective freedom in the exercise of the decision-making powers that are sovereign at any important social decision-choice period? What is the optimal size of the government to execute the program of the national interest and social vision and to implement the established conditions in the legal structure? What is the optimal constitutional apparatus required to ensure non-arbitrariness in the exercise of the state power to ensure fairness, justice and balance of freedoms for individuals and the collective?

In the modern politico-economic institutional arrangements where financial and real powers are concentrated in the hands of mega private concerns of production and distribution of income, how does the individual laborer get protection from freedom of abuse by those powerful concerns that have economic power to reduce individual life to nothingness and create conditions of indirect slavery? How does the social system design constraints on the possible arbitrary use of state power through the conduct of institutions of information classification on the basis of national security and public safety? What are the relationships among the economic, political and legal markets and to what extent do these relationships affect the structure of democracy, individual freedom and collective freedom? Should the markets, freedoms and decision-making powers be regulated and who should do the regulations? What kind of institutions must be established to prevent the possible reduction of individual decision-making power to just a phantom power through a complex electoral process for rubber stamping of the actions of powerful minority that controls the government for governance with the capacity to create unjust laws with the power of enforcement? What is the optimal power that must be invested in the government and how does the principal prevent an unnecessary expansion of this power by the agent for their own interest? These are important questions for the economic theory of social power and distribution that undoubtedly affect the distribution of any social cost-benefit configuration. The answers to them require critical analysis of the institutional relationship among the three structures of economics, politics and law.

It is within the social power distribution in the institutional trinity of the society that is defined the meaning of the notion that injustice is routed in the economic structure but the solution is defined in the political structure which can be used to refine and reposition the legal structure, as the social vision, national interest and social values change in the political economy. The connecting analytical point is that the reasonable understanding and social policy construct require a unified analysis of economics, law and political science. The current sharp separation in theory and studies of these areas is a limitation in the epistemological space for human decision-choice actions. In finding answers to the questions raised, it is a welcome positive direction for the increasing inter-disciplinary studies. These interdisciplinary studies will increase the understanding of cost-benefit externalities in policy constructs, institution building, social goal-objective formation, nation building and management of the institutional activities in the social set-up. They will further help the understanding of the social dynamics with an effective coordination of the economic, legal and political structures which undoubtedly will affect the conditions of scientific research and the progress of science. Confusions in the understanding of the behavior of applied and theoretical institutions are connected to income distribution, rent-seeking and profit enhancement as viewed in the cost-benefit space. They are also connected to the transformations of societies through qualitative equations of motion that help to change the social characteristics.

Finally, the problem of an optimal size of government seems to be a phantom problem because the government resides in good-evil duality whose size is constantly changing in response to the changing social values, preferences and moral tolerance. The government with any defined size is just like the *asantrofi-anoma* problem. It has unsatisfactory, temporary and unstable solution within states and among states in the cost-benefit duality [KKD] [KKD]. The concept of small-big government lives in polarity as well as in transformational unity with continual categorial conversions without end, which affect the nature of private-public-sector provisions of goods and services with continual power shifts for changing incentives, rent-seeking activities and profit enhancements. Every solution or resolution of the relative private-public-sector provision of goods and services or small-big government conflict in any form of democratic capitalism is a transformation of institutional arrangements that presents new conditions and a new set of problems. In this respect, the *solution path* is also the *problem path* to define the problem-solution duality in static and dynamic forms which connect the past to the present and the present to the future in the basic characteristic structure of the *sankofa-anoma* problem reflecting the problem of individual-collective duality which is neatly and compactly described by *anoma-kokone-kone* problem of the Akans in Ghana where the individual is both the solution and the problem to the society and the society is both the solution and the problem to the individual (see the Adinkra symbolism) [KKD].

VI: Defective-Deceptive Information Structures and Uncertainties in Democratic Collective Decision-Choice Systems

The political economy has been viewed as a decision-making system composed of power system of the individual and collective sovereignties which constraint the nature of the decision-choice variables in all the structures. Implicit in the decision-making system is a given social information structure. A basic reflection on the nature of the information structure will help to understand the analytical direction of the monograph and the use of toolbox of the fuzzy paradigm. The social information structure is composed of *defective information structure* and *deceptive information structure* which are used as inputs for the organic analysis in the monograph. They are distinguished by their sources, roles and characteristics in decision-choice systems. The defective information structure is general to all decision-choice actions of humanistic and non-humanistic systems. It is made up of *fuzzy information* and *stochastic information*.

The fuzzy information structure relates to language vagueness and representation, and ambiguities in transmission signals and thought. The fuzzy information is associated with quality of the general social information given the quantity of information. It generates *fuzzy uncertainty* and *fuzzy risk* which are associated with *possibilistic belief* in the decision-choice system. The stochastic information structure relates to information incompleteness, representation, transmission signals and thought in terms of volume. The incomplete information is associated with quantity of the general social information given the quality of the information. It generates *stochastic uncertainty* and *stochastic risk* which are associated with *probabilistic belief* in the decision-choice system. The defective information structure is the sum of fuzzy and stochastic information. The total uncertainty in the decision-choice system is the sum of fuzzy and stochastic uncertainties. The total risk in the decision-choice system is the sum of fuzzy and stochastic risks whose total belief system is the sum of possibilistic and probabilistic beliefs. The defective information structure involving vague and incomplete information is common to all knowledge areas.

The *deceptive information structure* is specifically associated with collective decision-choice system of cognitive agents. It is made up of *disinformation* and *misinformation* characteristics. It relates to language vagueness and representation, and ambiguities in transmission signals through intentional information manipulations to change the true information input for thought processing and direct individual and collective preferences to the desires of the manipulators. Its effect is more pronounced in democratic collective decision-choice system under majoratarian principle of one-person-one vote. The disinformation is a strategy to empty the mind of the decision-choice agents of what is known to be knowledge and to create *cognitive emptiness* relative a the decision-choice variable. The misinformation is a strategy to fill the cognitive emptiness of the decision-choice agents with information that may be made up of combined distorted signals to create a faked information into the social knowledge-construction in support of decision-choice actions.

Both disinformation and misinformation components of the deceptive information structure are associated with quality and quantity of the general social information given the defective information structure. The disinformation reduces the quantity of information input of knowledge which is constructed by a cognitive agent thus amplifying the stochastic uncertainty and the corresponding stochastic risk of the democratic collective decision-choice system. In this respect, all information classification and restrictions for reasons of national security, public safety, product advertizing and others belong to the area of disinformation structure which in turn reduces the integrity of the democratic decision-choice system. As such, they create lack of transparency and place the decision-choice agents at a disadvantage forcing them into an information sub-optimality in the collective decision-choice space. The misinformation reduces the quality of information thus enlarging the domain of fuzzy uncertainty and the corresponding fuzzy risk in the democratic collective decision-choice system. Propaganda of all kinds and false advertizing belong to the domain of misinformation. All trade secretes deprives the political economy of some quantity information which then affect the organic social decision.

The general purpose of deceptive information structure is to create cognitive illusions in the minds of decision-choice agents through the development of information asymmetry in the democratic collective decision-choice system where all sovereignties equally count. The effects of both the defective and deceptive information structures on the democratic collective decision-choice system must be examined and related to the political controls of the instruments of government for rent-creation, rent-preservation, rent-harvesting, the nature of governance, accountability and responsibility in the social set-up. This cognitive action will allow the establishment of a framework to critically examine the positive and negative impacts of information on the transformational conditions in the principal-agent duality, public-private sector duality, individual-community duality and the income-power distribution games within the costs-benefit duality. Furthermore, it will allow for the examination of the governmental direct and indirect control mechanisms of individual sovereignties in the democratic decision-choice system to create manufactured consents. Similarly, it will allow for an examination of information manipulation by powerful concerns to change the nature of objects under decision-choice action.

The analytical relationship here is that information control leads to power control and the effective control of the democratic collective decision-choice system that determines the national interest, social vision and the formation of the supporting social goal-objective set which then controls the individual and collective decision problems of the world in which we live and the solutions that may be required by theoretical and empirical analysis. In this respect, the task of this monograph will benefit from the counsel of Professor Joan Robinson who states: *The economics of equilibrium is a Moloch to which generations of students are still being sacrificed. I hope that I have been able to rescue a few here and there, not in order to offer them an easy life but to appeal for their help in the serious task of developing an analysis that deals with the economic problems of the world in which we live* [R2.69, p. xii-xiii]. In the real world that we live in, the problem path is also the solution path. Similarly, in the political economy the decision-making path is also the power-distribution path, all of which may be viewed in dualities and conflicts in continuum that provides the analytical strength for the techniques and methods used here. Among the phenomena of duality and conflicts are supply-demand dualities in all markets, creative-destruction dualities in all transformations, static-dynamic dualities in behaviors, exact-inexact dualities in language, true-false dualities in judgments, good-evil dualities in ethics, negative-positive dualities in energy systems and all categories of the principles of opposite.

VII: Cognitive Imbesility, Intellectual Zombism and Economic Reasoning

This section is a simple reflection on the statement by Professor Joan Robinson, quoted in Section VI in relation to economic thought and reasoning. The statement viewed in the practice of economic thought may be related to Keynes' reflection on the role of theories of economics in the policy and action.

> *The theory of economics does not furnish a body of settled conclusions immediately applicable to policy. It is a method rather than a doctrine, an apparatus of the mind, a technique of thinking, which helps its possessor to draw correct conclusions* [John Maynard Keynes]

In the development of economic theory and the applications that may be required of it, *cognitive imbecility* and *intellectual zombism* may arise at different levels of thought and application at different points of time. At the level of development of theories in economics, cognitive imbecility arise due to a failure to understand that knowledge development is a never-ending self-correcting process that is generated by cognitive forces under tension in either the same epistemological environment or different epistemological environment, where the future ideas seek to dislodge the present ideas, and the present ideas struggle to maintain their existence. When ideas emerge on the path of knowing, they are sheltered and maintained by the proponents through the development of protective walls of ideology of the paradigm of their creation, where such paradigm defines the environment of the validity of the ideas.

The different environments are defined, over the epistemological space, by the assumed different information structures which become inputs into the logical processing machine with a paradigm to create outputs that constitute the internal structures of the theories, conclusions and claims of truth. The information structure contains specific assumptions, vocabularies and grammar that are used to specify the boundaries of conclusions and claims of validity. The acts of holding on to, and repetitions of conclusions and logical claims of a theory without critical examination are acts of living in the zone of cognitive imbecility. Cognitive imbecility is thus a state of intellectual fixity on concepts and ideas at the level of theoretical development where these ideas are not critically examined for their validities. The individual does not develop the understanding that justifies the acceptance and repetition of the claims. In other words, these ideas are not internalized but simply mimicry. Such a state of thinking is referred to as the *intellectual state of cognitive imbecility*. Training to become an economist is simply training to repeat dogma and traditional ideas whether these ideas are relevant or not. For example in the graduate school student are train to study and mimic some mathematics, such as optimal control without being shown the economic questions that they can help us to answer. Fundamental questions of decision-making that affect the social system quality and the underlying institutional arrangements and development are swept aside. At the level of decision-choice practices, the state of cognitive imbecility becomes cognitively transformed into an intellectual zombie state where the socially irrelevant ideas are taken as knowledge which becomes an input into the decision-choice process. The decision-choice agents practicing in the zombie states just follow the dicta of intertemporal dead-ideas which are maintained in the state of cognitive imbecility even though the ideas are no longer relevant and useful for social progress. These dead ideas are ideologically useful in maintaining distortions in the social cost-benefit distribution.

Some of the ideas which are locked in the state of cognitive imbecility includes efficient markets for the construct of the individual and collective decisions, given the complexities social preferences, exact information whether complete or incomplete, the implicit assumption of a given institutions of capitalist type that fits and must fit all social formations on the principle of individual freedom. Another idea that is held on and constantly repeated is simply the private sector is more efficient than the public sector and that an allocation decision on the principle of individual freedom is more efficient than an allocation decision on the principle of collective freedom. Implied in these dead ideas is the notion that there is a government size that fits all social systems. The fact is that there cannot be a specified size of government that holds at all times and over generations for an analytic claim of small or big government. A permanent solution to the problem of well-defined size of the government is a phantom problem. The size of the government as measured in terms of the provision of goods and services relative to the size of the private sector is an unstable one which depends on ideology and intertemporal shifting of individual and collective preferences. The determination of the size of the government depends on the conflicts in the individual preferences in the collective decision-choice space. Within the same line of reasoning, the debate on regulation-deregulation phenomenon cannot be settled since it is intimately connected to the social responsibility of the institution of government and by logical extension to governance. We can speak of a regulation regime at any point of time but can we speak of an optimal regulation regime that can hold at all times and over generations.

The epistemic problem, here, is that sometimes the experts and authorities of the subject area have never been critically aware of the construction structure of the theories that they hold. They are sometimes not knowledgeable of the origins of the claims of ideas and the questions implied in the theoretical constructs. Furthermore, the defining characteristics and the meaning of the fundamental concepts that provide the vocabulary and grammar of the conceptual system and the uses of the paradigm of thought are either at reasonable distance or at variance, within the experts and authorities, which robs them of the foundational epistemic structure. They become imprisoned in the walls of the familiar, in a sense, imprisoned in a locked logical box. This cognitive imprisonment seal off logical possibilities as well as imprison epistemic creativity in thinking. Disagreements among experts and authorities in any area of knowledge are useful. They allow the questioning of the foundations and the conclusions that are obtained in any defined area of knowledge. They bring into focus the drive to restructure and redefine the foundation that involves the fundamental vocabulary and the grammar under the practice of the methodological principle of doubt. Some difficulties tend to arise when some persons seek more consistent and reasonably universal answers to escape the traps of the zones of cognitive imbecility and intellectual zombism. These persons, either from within or outside the class of experts and authorities, run into the frame where the needed answers require endless pursuit along the linkage between philosophy and what is claimed to be economic science. This is also true in other areas of knowledge production, but particularly disturbing in social sciences where ideological boundaries are substantially rigid and unbending. For example, the use of classical logic and it mathematics in economic analysis requires the fundamental assumption of exactness of the information structure and its representation. The insistence of the use of this fundamental assumption forces the experts and authorities to concentrate on the problems of quantitative disposition to the complete neglect of the qualitative disposition that involves subjectivity, vagueness and imprecision. This was the problem of Professors Joan Robinson [R15.40] and Karl Neibyl [R15.32].

The important point, here, is that a search to escape from an imprisonment within the walls of familiarity, as established by the past and present theories, requires sometimes a complete or partial rejection of the established and accepted positions of economic theories which are guarded by an established ideology. There are always conflicts in the present-future dualities in theoretical constructs, where the future seeks to violently dislodge the present by the methodological reductionism followed by methodological constructionism, and the present fights to defend and preserve itself in ideas and knowledge claims and resists the methodological reductionism that challenges the established ideas. Additionally, there is always a war against methodological doubt that seeks to restructure the epistemic foundations. In economics, these conflicts are amplified by ideological positions that over-cloud rational economic reasoning in both static and dynamic processes. The struggle to preserve outdated ideas is such that the members of the establishment of these ideas rejects and seek to destroy any idea that does not support, defend and preserve the outdated ideas. This is how states of intellectual zombism and cognitive imbecility are created and maintained by rigid publication system that is hostile to the practice of methodological doubt and new ideas that are not generated from within the walls of familiar.

At the level of university education, graduate students are encouraged to mimic the elements of familiarity containing some dead ideas, and they are rewarded for finding empirical support to maintain the acceptance of these dead ideas and then justify them on the basis of implicit ideology rather than a critical thinking. Critical thinking is encourage to the extent to which it seeks to preserve the established positions and help to build protective belts around them. In economics, these ideas are sometimes sheltered in classical abstract mathematics that have no relevance to the study of economics as part of the set of decision-choice problems of human behavior under conflicts, where these conflicts are defined on categories of dualities such as public-interest protection and suspension of constitutional liberties. How these conflicts, within the categories of dualities, affect allocation efficiencies and the resource space and distribution efficiency in the social cost-benefit space are practically neglected for the lack of quantitative data. For example, the economic-developmental processes are studied and analyzed in term of quantitative dynamics, where underdevelopment is explained in terms a dead idea of *vicious circle of poverty* based on income-savings-investment dynamics. Development as a qualitative process that affects the quantitative disposition under institutional tension is completely overlooked [**R15.4**]. It is the understanding of the implied qualitative dynamics for social transformations that empowers the analytical and scientific strengths of Marx and Schumpeter. In economics, the ideas within the walls of familiar are indirectly supported by political and legal principles under accepted belief system, where the proponents of new ideas of understanding, contrary to the established social order, are marginalized, prosecuted, intellectually ridiculed and sometimes jailed within the dark wall of prison of punishment. At graduate education, what must be thought is a critical thinking with the unshakable belief that the acquisition of knowledge is a process that resides in constructionism-reductionism duality with step-by-step movements as self-correction dynamics with unquestionable cognitive difficulties over the epistemological space. This critical thinking must be support with the principle of doubt in that what passes as knowledge at any given point of time may be a mistake. The principle of doubt must be supported with the principle of hope through epistemic industry in the defined hope-doubt duality in cognition. The hope-doubt duality is a logical mapping onto the constructionim-reductionism duality which is a logical mapping onto the actual-potential duality in the knowing process.

Generally, and in the practice space of decision-choice actions, any serious organic economic theory must deal with the multi-criteria, multi-objective and multi-personal problems, the solutions of which provide a temporary social equilibrium at a point in time, and permanent social disequilibrium over the decision-choice enveloping path where institutional arrangements tend to change to alter allocation and distribution behavior in the social resource and cost-benefit spaces. It is this temporary disequilibrium and the permanency of disequilibrium dynamics that led Professor Joan Robinson to state that *the economics of equilibrium is a Moloch to which generations of students are still being sacrificed.* Over the enveloping path, the distribution of social weights attached to the elements of the social goal-objective set, elements of the set of criteria and the elements of the set of personal net cost-benefits tend to change. Here, a critical examination is required for the relational-structural dynamics within the individual-collective duality and principal-agent duality in reference to the government-governance structure. This critical thinking requires a need to reconcile the held principles of the held duties of the agent to maintain individual freedom-rights structure of the members of the principal, and simultaneously protect the collective freedom-rights structure of the collective.

In general, the avoidance of traps of the states of cognitive imbecility and intellectual zombism requires the cultivation of both the methodological doubt and methodological hope with the view that knowledge production is a self-correction process [R7.14][R15.13][R15.14]. The self-correction process is stopped at the development of intellectual zombism and cognitive imbecility. This self-correcting process can be restored by the practice of both the methodological doubt and methodological hope. The practice of the methodological doubt through reductionism must lead to a critical questioning of propositions ideas and conclusions contained in any conceptual system including its foundational assumptions, its information representation, its approach to reasoning, and its laws of thought. It must also lead to a critical questioning of the nature and sources of information as an input into the knowledge construction process and whether the information structure is compatible with the phenomenon and consistent with the paradigm of thought. The practice of the methodological hope through constructionism must lead to a critical reconstruction of new ideas and conclusions that must replace the old ideas and conclusions contained in old conceptual system. This will include the foundational assumptions, information representation, approach to reasoning, and its laws of thought wherever they are needed. Similarly, it must also lead to a critical questioning of the nature and sources of information as an input into the knowledge production process and whether the information structure is compatible with the phenomenon and consistent with the paradigm of thought. It is the practice of both methodological doubt and methodological hope that forms the foundations of the development of this monograph.

The practice of methodological doubt finds expressions in the rejection of the classical assumption of exact information structure whether the information is complete or incomplete in nature and hence exact information representation. The classical paradigm with dualism and principle of excluded middle is rejected. The classical conclusions with stochastic conditionality is partially rejected and modified. The classical exact probability analysis is partially rejected and modified. The classical emphasis on quantitative analysis of the behavior in the quantity-quality dualism with the principle of excluded middle is rejected. The classical paradigm with its logic and mathematics is partially rejected and modified. The practice of methodology of hope find expression where the classical exact information structure is replaces with inexact information structure and representation wherever possible. The classical paradigm with dualism and principle of excluded middle is replaced with the fuzzy paradigm with duality and the principle of continuum. The classical paradigm with its logic and mathematics is replaced with the fuzzy paradigm with its logic and mathematics. The classical conclusions with stochastic conditionality are replaced with fuzzy conclusions with fuzzy-stochastic conditionality. The classical exact probability analysis is replaced with a fuzzy probability analysis. The classical emphasis on quantitative analysis of the behavior in the quantity-quality dualism with the principle of excluded middle is replaced with a simultaneous analysis of quality-quantity interaction within the quality-quantity duality with the principle of continuum which allows relational interactions of the behavior within the decision-choice space in the individual-community duality that generates forces of tension and changes in the social institutional arrangements and transformations to give meaning to social development under tension-disequilibrium and stability-equilibrium processes.

Contents

References: Interdisplanary

Chapter 1
Hermeneutics of Rent-Seeking Theory, Social Goal-Objective Formation and Fuzzy-Stochastic Rationality

In the monograph devoted to examining the political economy of social goal-objective formation in a democratic decision-choice system [R15.18], we discussed the problem of coalition formations as they relate to the political economy of collective decision-choice actions on national interest, social vision and the formation of their supporting social goal-objective set, on the basis of citizens' sovereignty. The concept of citizens' sovereignty relates to the distribution of the decision-making power. The goal-objective set is related to the decision-choice action on private-public sector combinations of the political economy that will show the directions of disappearances of outmoded institutions and the emergence of new institutions in terms of destruction- construction duality. The decision-choice process in the social decision-choice space led us to discuss the structure of coalition formations in the game of the collective choice of an institutional structure that may come to influence the dynamics of the social goal-objective formation.

The institutional structure of the political economy is seen in terms of public-private-sector proportions, irrespective of the measurements used. The institutional configuration is further seen in terms of the private-public-sector duality in a framework of analytics where every private-sector proportion has a public-sector proportion and vice versa in organizational unity. For the discussion on the nature of the private-public sector frontier and the collective decision-choice actions, see the works in [R12.4][R15.18]. This private-public sector duality in organizational unity is then related to the provision of goods and services in the social set-up. It is this public-private-sector proportional distribution of social production in the possibility space that tends to influence the decision-making process of the national goal-objective set, given the national interest and social vision. The organic structure of the decision-choice system in any modern social formation may be seen as composing of an interconnected sequence of 1) decisions on national interest and social vision, followed by 2) decisions on goal-objective formation in support of national interest and social vision, and 3) decisions on private-public sector combination for the provision and distribution of goods and services to accomplish the elements in the goal-objective set in support of the national interest and social vision. Each of these decisions involves sovereignty questions in terms of who has the power to make social decision for the collective. This leads us to the problem of the *power distribution* in the collective decision-choice space in resolving the problem of differences in the individual preferences in the collective decision-choice space. The approach to the construct of the power distribution may be framed as a decision-choice game.

The framework of the decision-choice game is established by creating an institution called the government or governmental system in terms of principal-agent duality. This is due to the cumbersome nature of involving every citizen in the social decision-choice actions. The institution of government is a natural instrument that is used by different groups called the social decision-making core, whose constitution is agreed to by the rules of law or customs or both. The decision-choice game in a democratic collective decision-choice space is set in two stages. The first stage is where the citizens assume the role of the agent and establish the social decision-making core as the principal. The second stage is where the social decision-making core assumes the role as agent and the citizens assume the role as the principal. The general structure of efficient democratic collective decision-choice systems is that the members in the class of the agents make decisions according to their preferences subject to the constraints of the collective preferences of the members of the principal under a set of agreed rules of behavior. To deal with the decision-choice game which is basically a voting game under a given information structure and a system of preference ordering, we introduced the concept of the social decision-making core $\mathbb{D}$ that constitutes the class of elected members from different political parties in the general population. The set $\mathbb{D}$ is partitioned into public and private sector ideologues and the independents of $\mathbb{D}_\Pi$, $\mathbb{D}_\mathbb{P}$ and $\mathbb{D}_\mathbb{I}$, where Π, $\mathbb{P}$ and $\mathbb{I}$ indicate public, private sector ideologues and independents respectively. Furthermore, we shall introduce the principal-agent duality with a continuum to present the structural dynamics of the democratic collective decision-choice process in the continuum.

To the ideologues, the central question on deciding on any goal-objective element is whether that goal objective element goes to enlarge either the private or the public sector. Those goal-objective elements that enhance the size of the public-sector (private sector) will be supported by the public sector ideologues (private-sector ideologues) without critical reflections on the overall welfare of the political economy. The decision-making process is such that the members of the independent class decide on the basis of the social welfare implications of each goal-objective element, and cast their votes on the principles of social cost-benefit rationality as they examine the elements relative to the best social welfare at a given information support. The size of the independent class becomes the dynamic force in the social decision making process to establish the public-private sector proportion that may come to rule. Such a proportion may come to shape the public-private sector institutional configuration and arrangements in the political economy. In fact, the preferences of the members of the independent class create the conversional moment that moves the system along the institutional efficiency frontier of the private-public sector proportions.

In this chapter, the social goal-objective formation, with a given set of institutional dynamics in the democratic decision-making that has been discussed in [R15.18], will be linked to the rent-seeking phenomenon in the political economy, where members of the non-decision-making core (the general public) as the principal, are allowed to develop strategies to influence the preferences of the members in the decision-making core as the agent. The central argument is that the formation of social goals and objectives is a game that is driven by the size of potential rent (net benefit), strategies of rent creation, abstraction of rent and protection of rent as payoffs for groups or individuals, if certain goal-objective elements are included in the goal-objective set. Thus, the creation of the social goal-objective set, national-interest determination and social vision definition are also economic production that obeys the laws of costs and benefits in terms of decision-choice actions. The outcomes of these three form the foundation on which social welfare is defined to relate to the efficiencies of resource allocation, output production and income-wealth distribution of the social set-up.

The supporting operating intelligence is the *cost-benefit rationality* which may be individual or social. An occasion also arises to present an interpretation of the rent-seeking theory of the politico-economic decision process under the *classical rationality*. The foundation of the logic of deregulation and prescriptions to governments as to how to mediate between the private and public sectors are abstracted from the theory as applied to political markets. It is argued that the government, as the agent in the social process of the principal-agent duality, is an institution of management and control that harbors various forms of potential non-wealth creating rents. The presence of potential rent is not market-regulation specific; rather rents involve the whole system of governmental social management and control including non-productive investments such as military production and some social-safety production. As such, any justification of market deregulation also applies to non-market deregulation and government outsourcing. The government outsourcing implies both privatization and deregulation of all economic and non-economic markets. The government outsourcing is an indirect privatization, and the direct and indirect privatization is to increase the size of the private sector relative to the public sector.

1.1 Hermeneutics of Rent-Seeking Theory

The formation of any social goal-objective set, as a democratic collective decision-making tool in reconciling individual conflicting preferences, is a multidimensional game with strategies and counter-strategies on the basis of social cost-benefit distribution. These strategies and counter strategies involve many opposing and social groups. The essential members of the game in the social set-up, after the decision-making core has been formed, have been aggregated into three opposing groups with common interests and similar social preferences. They are the social decision-making core $\mathbb{D}$, the public-sector-interest advocates, Π, and the private-sector-interest advocates $\mathbb{P}$. The public and private sector interest advocates constitute the members of the non-decision making core as well as the principal. The members of the social decision-making core, whose composition is under periodic revision, collectively constitute the agent in a democratic collective decision-choice system of the political economy. The structure of this particular principal-agent relationship is that while all the members of the agent are also members of the principal, most of the members of the principal are not members of the agent even though they all collectively own the institutional instrument of government that is created to be used to mediate individual and group conflicts in the social space. The relational structure within the principal-agent duality with a continuum and the power distribution associated with it constitute the essential foundation of the democratic collective decision-choice system. The principle of continuum in the principal-agent duality preserves the continuity where some members in the principal may enter as part of the agent while some members in the agent may enter as part of the principal in accordance with the politico-legal rules of the decision-choice game.

There are some theoretical and applied difficulties in this decision-choice structure, as well as there are essential problems to be solved. The difficulties involve the idea that the agent, as a representative government, is composed of individuals in the social decision-making core that have their own preferences over the elements of a social goal-objective set, in relation to the private-public sector combinations in the provision of social goods and services. Such preferences may be interfered and inspired by personal ambitions and net benefits that may have negative or positive impact on the welfare of the society. In fact, the government may be used by the social decision-making core to effect fairness or unfairness in the mediational process within the justice-injustice duality in the political economy. The majority of the members in the agent may hold a vision of the society that is completely different from the majority of the members of the principal. The principal, as the governed, is composed of members of the public that have their own individual preferences over the elements of the goal-objective set, in relation to the private-public sector combinations and other vision of the society that may be different from the agent. What the majority of the members in the principal may consider as the national interest may also be different from what the majority of the members of the agent may consider as the national interest.

The first essential problem is how to reconcile the conflicting and competing preferences of the members of the agent and the members of the principal after their decision to establish the agent (the representative government). The second essential problem that must be solved within the political economy is how the members in the non-decision-making core influence the members in the decision-making core to act in accordance with the preferences of the principal under democratic collective decision-making and majoritarian principle without violence. There is a third difficulty that involves the problem of asymmetry in information and power distribution in the social set-up. The presence of asymmetry of information and power always works to the disadvantage of the members in the non-decision-making core (general population), and in favor of the members of the decision-making core who control the political structure and hence the legal structure. This asymmetric information-knowledge structure may be used, and usually used, to alter the cognitive calculus of some members in the non-decision-making core and place them at sub-optimal decision-choice positions relative to the ones that would have been reached with the true information-knowledge structure.

The asymmetric information structure, on the part of the members of the principal, involves defective information composed of the characteristics of limitativeness and fuzziness in addition to deceptive information structure composed of the characteristics of disinformation and misinformation. This is where the need for transparency and truthfulness of the ruling decision-making core becomes important issues in governance and the management of the social setup. It is also through the lack of transparency and truthfulness that governments, as an institution, can be hijacked by a group to carry out an agenda that is benefit-group specific, but not beneficial to the welfare of the general society. It is here that electoral process in electing democratic governance may lead to the election of dictatorial governance. The information asymmetry may create political deceptions where such political deceptions in the democratic collective decision-choice space will continue as long as any membership in the decision-making core is not threatened. The only effectively available tools to the members of the principal class in a democratic collective decision-choice system are legal or illegal persuasions or both through an influence or a threat of expulsion through the exercise of voting or other legal means. The phenomenon of deception through calculated instruments to change the collective preferences of the principal by the members of the agent leads to the creation of a potential environment for violence, social revolt and social instabilities with the corresponding repression of the members of the principal by the members of the agent.

The practice of the democratic collective decision-choice system deviates from the ideal where fairness is to be balanced by governmental mediation in the spectrum of the justice-injustice duality. In practice, this ideal is altered into critical conflict in the democratic collective decision-choice space whose basic structure of the political decision-choice game involves the use of social goals and objectives through the legislative process to create and protect rent-seeking environments with wealth transfers. The process requires the members of the principal to actively work to include in the social goal-objective set, the goals and objectives that create rent-seeking environments for social groups where specific rents may be exploited and harvested by specific groups. In this way, the members of the decision-making core (the agent) or the elected officials become nothing but economic puppets to non-politically elected salespersons. The game is played with the constitution as the umpire. The whole process involves: 1) the establishment of the social decision-making core (the elected officials or the members of the agent), 2) the selling and buying of influence in order to affect the direction of votes to create the social goal-objective set, national interests and social vision, that define a rent-seeking environment favorable to specific interest groups, and 3) the exploitation of the environment for private gains through non-wealth-creating transfers through resource allocation for output production and income-wealth distibution. Number one is called influence tempering; number two is called rent-creation; and number three is called rent protection and harvesting. The structure of the selling and buying of influence is continually shifting on the basis of preferences that rotate the social welfare function on the private-public sector efficiency frontier [R15.18]. The structure of these three sequential steps has an important impact on the manner in which the members of the principal effectively participate in the democratic collective decision-choice system. It also has important effects on the generation of deceptive information structure and the form of the politico-economic game in the social space.

The political game is a continuous process where the winners create, exploit and protect elements in the social goal-objective set and national interest that lead to rent creation, exploitation and protection of possible rents in all markets. The social goal-objective formation, in trying to shape the welfare of the social setup, becomes a rent-production process that has inputs and outputs or costs and benefits for private exploitation of the public and private wealth accumulation. For analytical convenience, it is useful to view the rent-seeking game in the two aggregate sectors of a *real production system* and a *financial asset production system*. These two aggregate sectors may then be viewed in terms of transaction activities in the labor and goods markets of the real sector on one hand, and bond (security) and money markets of the financial sector on the other hand. The *political structure* offers opportunity for *rent creation* and the *legal structure* offers opportunities to *protect rent*, while the *economic structure* with the four aggregate markets offers opportunities to *harvest rent*. The relational form of the politico-economic environment for rent creation, protection and harvesting in the democratic collective decision-choice space is presented in Figure 1.1.

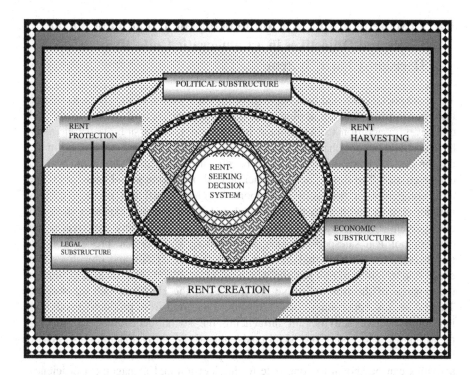

Fig. 1.1 A Cognitive Geometry of Relational Structure in Defining the Environment for Rent Creation, Protection and Harvesting in the Democratic Collective Decision-Choice System

The activities of rent abstraction in the economic space may help to improve output production and thus increase national wealth. They may also produce toxic substances that may help to reduce the size of the national wealth or to retard its growth through their negative effects on the efficiencies of the market, political and legal institutions through direct or indirect corruption. The process may also produce a situation where the preferences of a minority may come to define the social welfare configuration at the institutional frontier of private-public sector combinations as an optimal solution contrary to the principle of majority preferences. The system of the political economy, instead of being democratic by majoritarian principle, may degenerate into minority dictatorship where social preferences are established by an effective minority in terms of distorted social cost-benefit values.

1.2 Cost-Benefit Values of Influencing the Creation of Social Goals and Objectives in a Democratic Collective Decision-Choice System

There are social costs and benefits involved in the creation of the elements that go into making up the admissible and implementable set of social goals and objectives in addition to the national interest and social vision. These elements have cost and benefit characteristics over which preferences are established in the rent-production process. The costs and benefits are made up of direct and indirect characteristics. The direct social production cost of making social decisions is the total compensation that goes to support the elected officials in the decision-making core in addition to the incidental and miscellaneous expenses of social management and governance.   Besides these costs, the society experiences the costs of political salespersons such as lobbyists, who exercise pressure on the members of the decision-making core to select social goals and objectives that create opportunities of rent seeking which are favorable to some interest groups, at the expense of the true collective preferences of the electorate and the society. These costs are opportunity costs of resources diverted from productive activities into unproductive activities in influencing the social decision-choice process regarding the selection of the social goals, objectives and national interest.   There is no social utility from the activities of the political salespersons except, if these activities can be shown to contribute to the decision and management efficiency that can improve social productivity and welfare.   In this case, a social benefit can be established to allow net benefit to be examined for the utility of the political salespersons in the democratic collective decision-choice system. When positive net benefits are shown through the improvements of the social decision-choice process, such activities of the political sales-persons would be wealth-creating rent-seeking activities that go to improve the welfare of the social setup.

Similarly, there are opportunity costs to society in the production and establishment of special legislations that favor interest groups by the members of the decision-making core. It is important to observe that we have refrained from the use of the concept of government as the decision-making entity. Rather, we speak of the social decision-making core from the viewpoint of the logic of the activities of the social organism that has broad general application to all social formations in reconciling the conflicts in individual-community duality. The government is a permanent social institution of a holding place that houses the temporary decision-making core that varies from time to time. The government will always be there while the members of the decision-making core will vary over periods and generations as individual and social preferences shift over social goals, objectives and national interests. In general, this decision-making core may be composed of elected or non-elected members depending on the nature of the social formation, contract and governance. Within this thinking, the government is the people and the people constitute the government. The people are the backbone of the government. From the set of concepts defining social decision-choice systems, the government is just an entity; a decision-choice vehicle which is organizationally established and available to the decision-making core after it has been constituted by some general selection process that may include voting or any other means as designed by the members of the society.

The voting process, as an instrument of collective decision-choice process under the principle of majority rule, is simply the *voting game* in the cost-benefit space where the set of the rules of the game is part of the constitutional social mandates that establish more or less a general contract among the members of the society to transfer their individual social decision-making power to the members of the social decision-making unit called the decision-making core or the elected body, who come to inherit the government in a separate existence from the people, but in unity without violence. All the members of the voting public agree, under the constitutional contract, to temporarily surrender their rights of making social decisions (except their personal decisions) to the members of the decision-making core, to use the government as on organizational entity to serve the people, the public interest and the national interest through the establishment and implementation of the social goals and objectives, and to implement the social projects in order to reach the goals and objectives as designed toward the pursuance and maintenance of the national interest. Interestingly, the elements in the private and personal decision space are themselves under regulatory actions through the politico-legal structures. It is through the power invested in the government that motions of decision-choice elements take place between the private and public decision-choice spaces to establish changing conflicts in the individual preferences under cost-benefit rationality.

The services rendered by the decision-making core to the people, the public interest and the national interest are to achieve the best social welfare by using the political structure to manage the political and economic structures through the legal structure and the enforcement system. Given the political structure, the members of the decision-making core constitute the rule-law-making force to regulate the behaviors in these structures to create either both fairness and stability or both unfairness and instability in the social system composed of a multitude of competing individual preferences. To accomplish the task of the regulatory process, the government is equipped with the sovereignty to create instruments of enforcement and compliance. Generally, the social goals and objectives are created and implemented to provide transformations from one social state to the other. It is useful to see these social states as social *developmental categories* where each state is a category but a temporary one, subject to transformation through decision-choice actions. The implementation process and the social institutions, in terms of private-public sector configuration, provide the social system with a categorial moment that defines laws of transformation over the social states and in accordance with qualitative and quantitative equations of motion between developmental categories. When there is a categorial transformation, there is a cost-benefit distributional effect among different individuals and groups within the principal as a class. The distributional effects take place through institutional shifts and creations that are mandated by the decision-making core in the private or public sector or both. The social system as it is set up has a built-in complication of conflicts of interests for the members of the social decision-making core, where each member of the decision-making core is simultaneously a member of both the principal and the agent. The problem of the administrative process is how to minimize the conflict of interest to create fairness in the cost-benefit distribution in the social set up.

In the establishment and implementation of the elements in the social goal-objective set, conflicts arise between the individual and the collective on one hand and between the private and public interest on the other hand, as well as among the individuals in the goal-objective space. The conflicts are tensions that are reflected in the individual-community duality and generated by differential preferences in the corresponding cost-benefit duality in a continuum. Individual and group interests are specific in the social setup and invariably do run counter to each other and the aggregate public interest as seen in the cost-benefit distributional space. This is a classic duality problem in transformation dynamics in qualitative and quantitative processes [R4.57][R4.100] R7.14] [R15.13]. In the goal-objective space, there are constant solicitations of favors from, as well as exercises of pressure by interest groups, lobbyists, political action groups and political salespersons on the members of the decision-making core, to shape the preferences of the members in the decision-making core towards their respective preferences. The objective of the behaviors of these persons, here, is either to acquire influence or to indirectly control the political decision-making apparatus which is used to create social goals and objectives, legislate laws and regulations, design social programs and implement projects in a manner which will generate rent-benefit flows to the advantage of specific groups at the expense of the public, as well as cost flows that will fall on the general population. The whole process may be seen in terms of distribution of social costs and benefits in the socioeconomic production. It is here that the understanding of the theory of cost-benefit analysis acquires its scientific usefulness and the practice of the cost-benefit rationality also acquires an analytical potency in the general decision-choice system.

The activities of the political salespersons are the political-investment processes in the socioeconomic system. The political-investment process is to accumulate an influence capital for the acquisition of rent production. The success of this influence acquisition to the individual competing group will depend on the relative power held by the group. The dynamic effect of the political investment is to construct an influence-tampering force through direct and indirect gifts, monetary contributions and their equivalences, with *corrupt practices* that slowly destroy the democratic decision-choice system which consequently leads to the generation of the benefit of group-specific interests at the expense of general public welfare. In order for this process to continue, governmental transparency becomes a casualty where its place is taken over by deception and social revolving doors for public exploitation, injustice and unfairness. Here, capital accumulation and the investment process take place in the political market where inputs are private resources (costs) to generate positive net private benefit with cost-shifting to the public. The process, therefore, is cost shifting from those with greater influence and political capital to the weak and the poor in the democratic collective decision-choice system. In this way, democracy becomes a casualty to market mockery.

There are social costs and benefits associated with this process. The social cost of each group-specific objective that enters the social goal-objective set is the benefit of public welfare foregone if such a benefit is outweighed by the private cost of creating it. In this game, the odds are in favor of the more organized and affluent as well as the corporate sectors from the private and public sectors, and against the less affluent and poorly organized. The more organized is a group and more affluent that a group is the more the social power space that it is able to acquire for influence-tempering in shaping the preferences of the members of the decision-making core in the democratic decision-choice process. In this process, levels of education are useful to the extent to which they are used to acquire appropriate information-knowledge structure through collection and processing to strategize for the game. It must also be said that the greater the proportion of the members of the social decision-making core that belongs to the affluent and the best organized group, the greater is the unfairness and social corruption in the democratic collective decision-choice process. The corruption must be seen in terms of social cost-benefit distribution relative to social cost-benefit rationality [R3.1.4] [R3.1.23] [R3.1.24].

It is no longer a simple idea of exercising individual political right just to vote to create the decision-making core and hope that the public interest and social welfare will be served. This idea of equating voting with democracy becomes an ideological box with a complex number lock that holds the masses in the zone of cognitive imbecility, deception and citizens' irresponsibility in the collective decision-choice space. The right to vote has no decision importance after the process to establish the decision-making core has been completed, because all the voting individuals willingly agree to surrender their social decision-making rights to the elected members, and transfer the power of the state to them. The voters cannot recall their rights by the rules within the specified period and by specific process. An individual vote, even though it helps to establish the social decision-making core, does not provide access to the members of the decision-making core who must determine the national interest, social goals and objectives, and the social programs and policies that will support them in the structure of the private-public sector duality of the political economy. To gain access and power to play the game of national interest definition and social goal-objective setting, individuals must work and spend enormous resources to find people of like-mindedness that would allow an effective interest group to be formed. Alternatively, one must have substantial private resources to create and accumulate an *influence capital* that can be utilized on the members of the decision-making core who will decide on the elements that will enter the social goal-objective set in addition to the national interest and the budgetary distribution in supporting them. It is here that the formation of political coalitions discussed in [R15.18] becomes important, particularly for those whose personal resources are meager. Let us keep in mind that the democratic collective decision-choice system is composed of processes, transformations and rules of engagements. In the process, the individual right to vote is also a right to willingly surrender part of the individual freedom and transfer of sovereignty right that one has in the social decision-choice process. The right to vote on the part of the individual involves the right to establish character of the social decision-making core. The right to transfer sovereignty is also the right to empower the members of the social decision-making core to assume the position as agent who is charged with the responsibility to act on behalf of the general public. The morphology of the relational elements in the democratic collective decision-choice system is presented in a cognitive geometry in Figure 1.2.

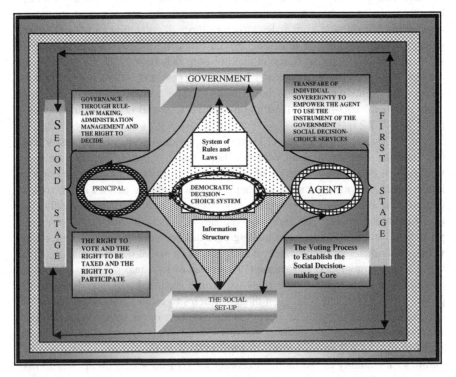

Fig. 1.2 A Cognitive Geometry of a Democratic Decision –Choice Systems showing the Relational Interactions of the process, the Participants, the Rule-law Structure of the Principal-Agent Duality

The social costs of the exercises to create the social goals and objectives are the resources used to lobby simply to influence the creation of the set of social goals, objectives and legislations that will alter the legal structure, which then will redefine the environment for the individual and social decision making in the political, economic and legal markets. In these markets, everything is viewed as a commodity for trade where supply and demand are the driving forces of cost-benefit conflicts in the democratic social setup based on the fundamental ethical principle of individualism. In order to assess whether or not the influence-acquisition and influence-tampering process is socially beneficial, we need to compute the total net social benefit associated with the process. In all accounts, it may be demonstrated that the social costs of lobbying and exerting group influence on the members of the decision-making core involve economically unproductive activities which do not lead to wealth creation, but rent-seeking opportunities for transfers of social wealth to particular individuals and groups. These activities are merely a social waste in terms of the rent-seeking argument of government operations [R13.3] [RI3.6].

1.2.1 Political Information, Propaganda and the Democratic Decision-Choice System

The sociopolitical game in the goal-objective formation on the basis of the democratic principle of majority rule works to establish itself on the optimal path through political information collection and information-processing as part of the social information by the individual participants. Information possession and information processing capacity are important ingredients in the efficient functioning of democratic collective decision-choice systems. Whoever controls both the social information and the corresponding decision-choice outcomes indirectly determines the national interest and the supporting goal-objective set. In fact, the information control belongs to the class of democratic destruction through indirect dictatorship. The information control is done by appropriate agencies with propaganda manipulations in the social information space to create a deceptive information structure composed of disinformation and misinformation.

The instruments to achieve the deceptive information structure include but are not limited to: 1) limiting the quantity of social information through a number of classificatory systems, and 2) changing the quality of information through disinformation and misinformation at the practice of social decision-choice actions in the political space. The individual and collective information-processing capacity is influenced through the institutions of education, their access, curriculum and the culture of thinking. The aggregate effect is simply to constantly introduce elements of vagueness and ambiguities into the social information space which must be defuzzified with an appropriate reasoning by the individual members of the society in accordance with their capacity to distinguish relevance from irrelevance in the use of information in the political decision-choice space. The defuzzification is equivalent to decoding the elements of the deceptive information structure which contains misinformation and disinformation items that are crafted into an ideology to shape the psychology of the social decision-choice process.

The psychology of decision-making is that, the social information, irrespective of the source, is taken to be knowledge of the surrounding social area of the national system. The national system is projected as the internal state of the social system and external conditions of threats and rewards which the controllers (decision-making core and sub-agents) employ in implementing the needed tactics and strategies of the social stability in the democratic collective decision-choice system that generates a self-controlling, self-organizing and self-correcting process through decision-choice actions, with implementations that involve the public-sector consciousness and private-sector consciousness in duality, continuum and living unity.

In terms of individual decision making, knowledge is cognitively abstracted from the available social information that people believe to possess, and facts which are external to them. The facts which are external to the individual and the collective assume the character of information signals in the process of transmission. This information transmission process is accomplished through the energy field where each information signal is a set in motion propelled by sufficiently corresponding energy without any loss of the essential characters of the information signal. The energy carrier of the information signal of the external data may be amplified or de-amplified for increasing the complexity of the deceptive information structure to affect the self-correcting element of the social decision-choice system. For the social information signal to retain the value corresponding to the fact or data external to the recipient, the amplification and de-amplification of the energy carrier must stay within permissible bounds which may be referred to as an interval with fuzzy covering, where every fuzzy covering is a fuzzy continuum accompanied by a corresponding *fuzzy **conditionality*** [R15.13].

The social information, being transmitted to the individuals or the collective, can be distorted by over amplification and under amplification of the energy as the force of transmission to alter the individual decision-choice responses to the collective decision-choice system. This is how propaganda is created to alter the information field that decision-choice agents operate in and within the democratic decision-choice system. The information field is projected onto the energy field for a continuous connection in the sending-receiving channels. At the energy field, disinformation is associated with energy in over-amplification while misinformation is associated with energy under amplification. The over and under amplifications are associated with propaganda that is intended to distort the receiver's information signals and affect the outcomes of the knowledge processing that enter into the decision-choice system of the individuals and the collective.

The use of information distortions to create deceptive information structure through propaganda is an important but indirect way to the destruction of democratic social formation and the democratic decision-choice system on the principle of majority rule by changing the information on cost-benefit imputations for the individual and the collective. Changing the cost-benefit imputations, relative to the individual preference ordering, changes the individual optimizing decision in the collective decision-choice space. This process distorts the true decision-choice intents of the individuals as well as the cardinality of the subsets of the votes in the political space. Such a process is nothing but an indirect way of vote rigging. It is not different from deceptive advertising to shift preferences in an economic space under a market system of output distribution in favor of different commodities.

Deceptive information structure through deliberate distortions, in the political information space, enters into the democratic collective decision-choice system through the use of the institutions of the government and mass media by the social decision-making core and the individuals seeking to enter the club of the decision-making core with a supporting set of individual potential voters. The government's distortions enter the democratic collective decision-choice system under a complicated management system of information classification through the concept of the organization of the national security as protection for the state against internal and external threats. In this process, the transparency in governance becomes a casualty at the deadly sword of a national security state through the strategic use of the disinformation-misinformation process to force the citizens into a suboptimal zone of democratic collective decision-choice actions. The disinformation-misinformation process create mirage of democracy, freedom, justice in the illusion space of decision-choice sovereignty in which the individuals operate.

Misinformation over-amplifies the allowable area of social decision-choice actions and pushes an individual in the collective into an upper illusory frontier of decision-choice action, forcing some individuals to select a social option which cannot be supported without the misinformation strategy. Disinformation, on the other hand, does the opposite. It under-amplifies the allowable area of social decision-choice actions and pushes the individuals in the collective into an illusory frontier of the social decision-choice action, forcing some individuals to select a social option which cannot be supported without the disinformation. The tool of disinformation is to strip from the individual mind what is known to be the truth and create cognitive emptiness. The tool of misinformation is to fill the cognitive empty space of an individual with vague ideas that may be made up of combinations of truth and falsity, thus creating illusions in preference ranking of social states under cost-benefit rationality.

The creation of deceptive information structure through disinformation and misinformation is to change the individual cost-benefit imputations over the social space in order to influence the individual voting action in the democratic collective decision-choice system. All these are related and must be referenced against the nature of the social cost-benefit output where the ultimate objective of the social system in its political decision-choice actions is to optimize the net social cost-benefit output. Relative to the outcome of the social net cost-benefit output, the ultimate objective of an individual participant in the democratic decision-choice system is to optimize his or her net cost-benefit through the distribution of the gains of the general social production at each point in time irrespective of the outcome of the private-public sector combination in the social set-up. The individual social net cost-benefit and the collective social net cost-benefit reside in dynamic dualities with continua, where, by manipulation of information, some members may stand to gain at the expense of others and the society. Deceptive information structure through the disinformation-misinformation process may distort the path to the social and individual ultimate decision.

So long as democratic decision-choice agents are ignorant to the effects of factors of disinformation-misinformation processes on the democratic decision-choice system, they become nothing but mechanical objects, under indirect manipulation. In this way, they are moved to the zone of powerless in relation to the active social decision actions and simply become rubber-stampers of the democratic decision-choice system under the control of the government, the decision-making core and the privileged class. Furthermore, as long as the democratic decision-choice agents underestimate the role that money and resources play in shaping the influence of deceptive information structure, they become victims to ideological propaganda rather than being active participants in the democratic decision-choice system in resolving conflicts in the individual preferences under the collective decision-choice space.

The more social information is restricted, by security classifications, to the general public in the three markets, the more is asymmetry created in the efficient exercise of the fundamental ethical principle of individualism and the more sub-rational decisions come to be associated with the individuals or groups in the democratic process, thus curtailing the power of the individual freedom within the social contract in the democratic collective decision-choice system. The effects of the information-knowledge asymmetry and wealth differentials do not only shape the individual performances in the political and the legal markets, but they become more pronounced in the economic markets, particularly, in economic sub-markets, such as financial economic sub-markets where games in expectations and risks are more information-knowledge driven [R15.1], [R15.14] [R15.34] [R15.37]. The markets where individual freedom is to be exercised to strengthen democracy assume the role of a mockery of democracy, where democracy simply becomes an ideology used to quiet discontent, rather than as a vehicle of resolving the conflicts in the individual-collective duality in the preference space within the collective decision-choice actions. In this respect, the practice of democracy, instead of expanding the space of liberty, destroys individual freedom and decision-choice sovereignty by generating a mockery through a complicated manipulation of deceptive information structure.

The efficient function of the democratic decision-choice system requires that the objective nature of social information constitutes a necessary attribute of the supporting information reliability and its knowledge output for individual decisions in the social decision-choice space, if the individuals are to be real active participants in the democratic system of decision-choice actions. The task of an individual in the democratic collective decision-choice system is to participate to create an efficient social decision-choice system under the conditions of non-deceptive information structure either by the members of the social decision-making core using the institution of the government, or by those who seek to re-enter into the social decision-making core. When the information structure is deceptive, it disorients and disorganizes the domain of the knowledge production in support of the social decisions of the individual democratic agent where market imputations in the three structures make a mockery of true democracy. This leads to the production of apathy and a superficial attitude with regard to one's responsibility toward the collective, the democratic collective decision-choice system, the government and the governance.

The essential ingredient for efficient individual participation in the democratic collective decision-choice system of a political economy organized on the principle of individual freedom and decision-choice sovereignty is the degree of truthfulness contained in the socio-political information structure in the social space. Such information truthfulness must reflect the actual conditions that correspond to the smoothed functioning of the individual participation in the democratic process on the basis of the fundamental postulate of individual freedom in the three structures. Deceptive information, in the political economy organized on the principle of individual freedom of decision-choice action, compromises the integrity of the democratic collective decision-choice system, where the strong and powerful benefit at the expense of the weak and poor who are ideologically made to believe that they are participating and contributing to the social conflict resolution under democracy. The individual freedom is to strengthen the collective freedom, but the compromise in the democratic social decision-choice system destroys the collective freedom that is essential for the self-correcting character of the social management, and distorts the cost-benefit imputations for the adjustment process in the favor of the producers of the deceptive information structure, especially in the decision-choice process to establish the social decision-making core that occupies the institutions and the house of the government.

The elements of the deceptive information structure are carried by the mass media, institutions of propaganda and supported by those who have money and resources to acquire the channels of information dissemination. In this process, the democratic collective decision-choice system operates on the principle of weighted democracy where the weights are defined by either money or resource. The intent of the deceptive information structure, at the level of the population, is to manipulate public opinion and shape preferences of individual decision-choice agents in favor of some elements in the social goal-objective set. This is done through the election of candidates, who will be favorable to the selection of these elements, into the decision-making core who are charged with the responsibility for exercising the power of the governmental machinery. In this respect, what to inform, when to inform, where to inform, who to inform, their effective organization and coordination of the informing are essential to the outcomes of the support by the members of the decision-making core for the elements in the social goal-objective set. But, the effective organization and coordination require money and resources which favor the preferences of the rich over the poor. We may add that the creation of a deceptive information structure and its implementation in the social information space corrupt the value structure of the social set-up, and distort the essence of the democratic decision-choice system against the preferences of the weak and poor, in favor of the preferences of the powerful and rich in the creation and maintenance of the elements in the goal-objective set and their impact on the distribution of the social cost-benefit configuration. The ultimate result is an effective cost-benefit shifting process, where cost is shifted disproportionately to the poor and benefit is also shifted disproportionately to the rich and powerful through the creation of a system of rent-seeking processes that generate social waste in the production system as the political economy moves over different private-public sector combinations. The implication of the analysis of the process of social goal-objective creation and national-interest determination being advanced here is the view that *social waste* is not limited to economic public monopolies and a system of government regulations as we are made to think by some authors in rent-seeking theories on government [R13][R13.5] as well as theories of economic behavior in a capitalist economic organization [R11.23] [R11.24] [R11.44][R11.51][R11.53] [R15.45].

Social waste is a characteristic of the process of social decision making on the basis of the establishment of the social decision-making core on the principles of some democratic rules and the fundamental ethical principle of individualism that contains conflicting individual interests and preferences. The social wastes may be viewed as the costs of social decision making under democratic regimes of governance, where the emphasis is placed on individual interests and individual freedom to choose as the driving force for determining the aggregate outcomes of the private-public sector combinations in the provision of goods and services which are consistent in supporting the national interest and social vision. There are conflicts between the public-good and private-good creations where the lobbying process seeks to transform the private interest into the public interests and private costs into social costs with the hope of creating rent opportunities for private harvesting. The power of the process is that individual net profits are privatized while individual net losses are indirectly socialized in terms of Schumpeterian socialism where the bureaucratic capitalist class directly or indirectly controls the decision-making power. The process of creating a rent-seeking environment is the *rent-seeking game* which requires *rent-seeking strategies*. After an optimal rent seeking strategy has been formed, it must be supported by a rent-protection game and strategy, after which it must be supported by rent-harvesting games and strategies.

At the level of the collective decision process, under a democratic collective decision-choice system involving the structure of the political economy, some questions of private-public sector decision-making come into interplay. What criteria should the individual and hence the collective use in determining the choice of public-private sector combinations in providing social goods and services? In this respect, should the criteria of public sector decisions concentrate on defense and national security in support of the private sector interest to the neglect of the interest of the collective? Alternatively, should the criteria of the public sector decisions reflect the weighted sum of full-employment, economic growth targets, incomes' policy (income distribution), and the balance of trade and payments that are seen in relation to the collective interests and freedom in the political economy? How should the public-private sector relationship be seen in the political economy under a democratic collective decision-choice system guided by the individual preferences as expressed by the fundamental principle of individual freedom and sovereignty in the social decision-choice space? These questions relate to the nature of private-public partnership in the political economy.

1.3   A Reflection on the Democratic Decision-Choice System in the Political Economy

The objective of social formations is to organize human and non-human resources to provide needs and wants for the members that constitute the resulting society or the political economy. The organization to satisfy the needs and wants requires the provision of goods and services. The organization of resources and the provision of goods and services demand an organization of a decision-choice system and the casting of institutions which may come and go depending on the cultural dynamics of the collective belief system. The organization of the decision-choice system requires the management of the individual and the collective decisions in terms of their relative freedoms and the distribution of such freedoms over the decision-choice elements which are relevant to the progress and stability of the political economy. Here, the decision-choice space must be divided into the individual and the collective spaces in terms of power and freedom to act on the decision-choice items. With the general decision-choice system of the political economy in balancing net needs and wants on one hand, and individual and collective conflicts on the other hand in the general decision-choice space of the political economy, emerges the individual-community duality that is translated into private-public sector duality with a continuum for continual social transformation.

The question then is: what set of decisions should be left to the private sector and which freedoms should be accorded to the private sector in carrying out its responsibility of the provision of goods and services in the political economy? The answer to this question requires a resolution to the problem of what goods and services must be produced in the private sector and public sector separately and public-private sector jointly. There are other questions that are related to the social decision-choice actions in the political economy. What relative freedoms should be accorded to the private and the collective and what relative responsibility should the private sector and the public sector hold to the social organism for an efficient functioning of the political economy? We must also find an answer to the question: what incentive structure must be designed to support the aggregate decision-choice processes of both the public and private sectors? Should the public subsidies to the private sector be part of the incentive structure? Should the government financing be part of the incentive structure? Should the development of the social infrastructure be left to the responsibility of the public sector or to the responsibility of the private sector or to both? These are classic questions that every social system must answer in the conflict spaces in the individual-community duality as well as the private-public sector polarity.

The answers to these questions depend on the conditions of three cases of the form of the social organism. The answers will also point to the nature of institutional configuration that must be created for the social setup. One form reflects a complete private sector for the provision of goods and services. The other form reflects a complete public sector for the provision of goods and services. The third form reflects a joint public-private sector combination for the provision of the goods and services to the society. The first form is what the economists call pure capitalism (pure capitalist political economy); the second form is pure communism (pure communist political economy) and the third form is called the mixed economy (mixed political economy). Corresponding to these three forms, emerges three different types of political economy with three different sets of combinations of economic structure, political structure and legal structure that together spin individual and collective behaviors in the social decision-choice space, as well as generate conflicts in the individual-community duality and continual restructuring of institutional arrangements in the political polarity.

These types of the political economy may be democratic, non-democratic or both in their social decision-choice structures. The criterion that is socially accepted to govern the individual and collective decision-choice processes may be different for each socio-political form, depending on what the members expect the social decision-choice system to accomplish in any given form of the political economy. The criterion appears as the organic criterion and a set of specific criteria that must be consistent with the organic criterion. The organic criterion must be consistent with as well as provide stability for the political economy in a way that defines the boundaries of the micro-criteria. The organic criterion constitutes the socio-political foundation for designing the incentive structure in support of the micro-criteria that will shape the directions of the collective and individual decision-choice behaviors. In pure capitalism without government controls, the organic criterion is the fundamental ethical principle of individual self-interest in pursuing freedom of self that must be translated into individual welfare optimization in terms of satisfaction or profit or both as the micro-criteria without regard to the welfare of others and the collective freedoms of the members, except when such freedoms enhance the individual welfare. The responsibility of productive enterprise is to pursue profits as well as increase profits. The responsibility of the individual is to pursue self-happiness. The responsibility of the collective is not to restrain the individual in the sense that the emphasis in the conflict space of the individual-community duality, within the political polarity, is placed on the individuals who define the community and where the members define the set, on the principle of each for himself or herself.

Here, the measure of social progress is *gross individual happiness* (GIH) in terms of how well is an individual doing in any social state within the political economy. In this ideological space, collective social progress is irrelevant to the design of the system's incentive structure. The exploitation of one to the benefit of the other is the implicit principle of behavior. The legal structure is constructed to incorporate this principle to allow for acceptable boundaries of unfairness and social acceptance of some degree of exploitation.

In the case of pure communism, the organic criterion is the fundamental ethical principle of collective self-interest in pursuing freedom for all that must be translated into collective welfare optimization in terms of collective satisfaction or collective progress or both as the micro-criteria with full regard to the individual welfare and freedoms of the members within the confines of the collective freedom. The responsibility of the collective is: each for all and all for each and to provide collective assistance to the individual to contribute to the collective endeavor. The responsibility of the individual is to pursue happiness within the collective happiness and the measure of social progress is *gross national happiness* (GNH). The optimization of the collective welfare is under the state where the responsibility of productive enterprise is the pursuance of human welfare but not profit for profit sake. In this respect, the organizational emphasis in the conflict space of the individual-community duality, within the political polarity, is on the community that defines the individual as an element of the set where the individual identity makes no sense outside the community.

In the case of the mixed political economy, the organic criterion is a weighted sum of the collective and individual behaviors on the fundamental principle of weighted balance between individual and collective self interests that seek to reconcile the conflicts of individual freedom within the collective freedom, and collective freedom as a constraint on the individual freedom. The micro-criteria are reflected in the division of the set of optimizing behaviors in the cost-benefit space within the organic social decision-choice space. The social decision-choice space is divided into the individual decision-choice space and the collective decision-choice space in the sense that the individual claims identity within the collective and the collective claims identity from the members. The responsibility of the individual is to pursue happiness within the boundaries of the divisions as set. The responsibility of the collective is to work also within the boundaries of the division. The individual and the collective decision-choice spaces function as constraints on one another to establish mutual support. The individual and the collective function in a relational give-and-take mode in the sense that the emphasis in the conflict space of the individual-community duality, within the political polarity, is shared between the individual who defines the identity of the community which gives identity to the individual. The members and the collective exist in mutual definitions in the mixed political economy. The individual pursues satisfaction and profit and the collective pursues welfare optimization within the social set-up. The organic principle is not each for herself or himself or each for all and all for each, but rather give-and-take relations of freedom. Here, the measure of social progress is a weighted value of the *gross individual happiness* (GIH) in terms of how well an individual is doing and the *gross national happiness* (GNH) in terms of how well the society is doing.

The optimizing behaviors in the three forms of the political economy must be defined in the cost-benefit space of the decision-choice system in the use of the available national resources where the criteria and constraints are defined in terms of the opportunity costs of the destruction-construction process of resource transformations for the dynamics in the qualitative-quantitative duality of the political economy within each pole. The behaviors within the quality-quantity duality point to the dynamics of social transformations and changing institutional arrangements. Such behaviors are decision-choice driven to create benefits and costs on the enveloping path of the outcomes of implementations. Each form of the political economy defines boundaries of allowable decision-choice behaviors that are consistent with the integrity and stability of its form. It also points to compatible institutions which are constructible to make the functioning of the social decision-choice system possible, which will incorporate the individual and the collective behaviors within the consumption-production duality for both the individual and collective (social) wants and needs in the political economy. Here, appears another important difference and conflict in the individual-community duality within the political polarity. The individual goal is self-development and the maintenance of personal integrity. The collective interest is an organizational development and a nation building which must account for the individuals as well as for all the members in the political economy irrespective of its form.

In a complete capitalist political economy, one may conceptually have capitalism with either a government or without a government in governance where the mode of production is in the hands of the private individuals or groups. The analytical thrust is the existence of organization-un-organization duality where government-no-government duality emerges from the organization as the dual in the political economy. The conditions of government emerge out of conditions of non-government where non-government is a primary category from which a government is a derived category on social transformations. The government is an institutional organization whose role is either collective determined or may have evolved over time and is agreed upon. The presence of government creates another collective decision-choice conflict in the social organism in terms of what role should be assigned to the government, and what relative decision-choice power should be invested in the government, and who should control this power. The presence of a government in a complete capitalist political economy is restricted to activities in support of individual freedom of choice to pursue properties and private interest through unrestricted freedom to choose in the economic structure, especially if the government is controlled by the capitalist class. The government is given the responsibility to manage the state power for stability of conflicts in the individual preferences over the economic structure. The nation building is in the hands of the private sector that controls the economic structure. The government is to control the political and the legal structures. In a complete capitalist political economy, the conflict of private-public sector duality is resolved in the favor of the private sector from the political structure and enforced from the legal sector. The conflict in the individual-community duality still exists but its stability and management are the responsibility of government. The efficiency of the management and the degree of responsibility in the social collectivity will always depend on who rides the institution of the government

In the case of a complete communistic political economy, the government controls both productive capacity and governance in terms of economic, political and legal structural foundations of the social formation, where the emphasis is on collective freedom to choose. The government in the communistic political economy is also an institutional organization whose role is either collectively determined or may have evolved through some means which is agreed upon. The presence of a government creates another collective decision-choice conflict in the social organism in terms of what role should be assigned to the individuals, what relative decision-choice power should be invested in the individual for self-identity, and how should the individuals be related to the social decision-choice power. There is the government with the state enterprises. The responsibility of the state enterprise is the pursuance of efficiency of the collective welfare optimization in the sense of the best use of the nation's resources to support the political economy relative to the national interest and social vision. The government assumes the role of nation building by developing institutions that will support the public provision of goods and services to the members of the society. The public enterprises assume the control of the economic structure while the government directly controls the political and legal structures and indirectly controls the economic structure. The welfares of the individuals and the collective are the responsibility of the government who is also charged with the responsibility of the security of the state. In complete communism the conflict of private-public sector duality is resolved in favor of the public sector from the political structure and enforced from the legal sector. The conflict in the individual-community duality still exists but its stability is the responsibility of government.

The form of the mixed political economy, composed of private and public sectors, is extremely complicated in terms of the distribution of freedoms over decision-choice actions between the collective and the individuals in the economic, political and legal structure of the political economy. Between the two extremes of complete private and complete public sectors, there is a continuum that defines an infinite set of private-public sector combinations from which a selection can be made [R15.18]. In this respect of a choice of an appropriate public-private sector combination of institution, freedoms and sovereignty, the responsibility and accountability of the government become complicated and complex. The complexity translates itself into a complex decision-choice system of the government in terms of the relative responsibility to the public-sector and private-sector welfares to ensure the overall best welfare of the political economy, in terms of supply and distribution of goods and services and the institutional stability of the social organism, that must establish a reasonable balance between individual and collective freedoms in support of the nation-building effort. The nation building includes but not limited to the development of the social infra-structure, responsibility of the tasks that are too large for private sector to undertake, the development of public institutions in the politico-legal structures and the control of the state power. This nation building becomes the responsibility of the government that represents socially collective actions where the implementation of the nation-building tasks depends on the character of the social decision-making core.

Given the mixed political economy, a number of questions arise: What goods and services should be assigned for the public sector provision? Should such goods and services be limited to defense and military while the provision of other goods and services become the responsibility of the private sector? Should the provision of services from the social-infrastructure in support of both the private and the public sector be left to the public sector or the private sector? Should there be joint public-private sector participation in the provision of goods and services in the mixed political economy? Should the government interfere in the private sector incentives through a system of subsidies? To what extent should the private sector be involved in the management of public institutions and the public sector be involved in the private institutions? Should the government be responsible in managing the overall functioning of the mixed political economy? If the government is not responsible for the management of the political economy, then how should the coordination between the private and public sectors be administered? If the government is charged with the responsibility of managing the mixed political economy, should such a responsibility involve designing rules and regulations to distribute the social responsibility of the provision of goods and services between the private and the public sector in such a way that the private and public sectors are inter-supportive? In this connection, what should be the principle of the incentive structure that must be established to produce and manage an *optimal private-public mix* in the mixed political economy to ensure optimal social welfare? Two further questions emerge: What is the shared responsibility between the private sector under private operations and the public sector under the government operations in the social system in relation to the collective social welfare optimization for the members of the mixed political economy? What private-public sector combination should be selected from the infinite set to ensure the welfare optimum of the mixed political economy? It may be pointed out that this optimal private-public mix will vary over different generations and different cultural systems.

Should the answers and decision-choice actions of these questions reflect the preferences of the individuals and the collective in some form of democratic decision-choice actions? The most reflective question among the questions is the question of the decision-choice action on the appropriate private-public sector mix and the criterion that must guide the respective decision-choice process. The selected criterion must relate to the incentive structure and efficient resource usage as well as the nature and quality of the political economy. Given any private-public sector mix, should the private sector decision-choice actions be guided by the same criterion as that of the public sector decision-choice actions? If the answer is affirmative, then this criterion will also be the criterion for the general decision-choice structure for the political economy and all its relevant institutions. If there are two different criteria for the two sectors, then the management of the social set-up must confront the solution to the problem of appropriate criteria for all the private-sector decision-choice actions as well as the appropriate criteria for all the public-sector decision-choice actions and how these different criteria can be combined to develop an aggregate criterion to examine the welfare efficiency of the mixed political economy in the use of its limited human and non-human resources. The private sector decision-choice actions will fall under the private-sector micro-decision choice system. The public-sector decision-choice actions will fall under the public-sector micro-decision-choice system. The two combine to generate macro-decision-choice system which is under the management of the governmental machinery regarding the efficiency of national resource utilization and provision of goods and services. The study of the private-sector micro-decision-choice system is what is called the microeconomics in economic science. The study of the public-sector micro-decision-choice system is the public economics that includes budgetary decisions.

Given the private-public-sector mix, should the private-sector decision-choice actions concentrate solely on the private-sector interest? Should such interest, when translated into the space of the provision of goods and services, reveal itself as what some analysts hold as the social responsibility of the private-sector business organization? Should such social responsibility be specified to include the pursuance of profits as well as increases of business profits in addition to advancing the private interest and the freedom of individual decision-choice actions in the three structures? These must be done irrespective of what the collective interest and freedom might be in the mixed political economy, where the markets mediate any conflicts in preferences without government intervention. Alternatively, should the private business organizations use of the provision of goods and services combine profit interest and social responsibility to the collective so that when translated into the space of provision of goods and services, the pursuance of profit and the private sector interest is constrained by the collective interest, fairness and social welfare within the framework of the legal structure? In this case should the production of goods and services by the private sector be constrained by social decisions on the production of non-public sector goods and services?

In the political economic form of private-public sector mix, what kind of criterion or criteria must guide the public-sector decision-choice actions in the provision of public-sector goods and services and the overall management of the national economy? Should the criterion be the efficient production of a system of social infrastructure in support of the efficient production of goods and services in the private sector? Here, the social infrastructure is conceived to include institutions of the legal and political structures, defense and national security in terms of internal and external system's integrity. The internal security includes the provision for services involving social safety nets due to distortions of the market solutions to resource allocation, product production, income distribution and environment for the individual and collective existence. The point of emphasis is that these problems are institutionally generated and their solutions are decision-choice determined, just as the structure of the institutional configuration is decision-choice determined. The success-failure outcomes of these complex decision-choice processes point to the direction of national history which is simply an enveloping of the success-failure decisions in the national history.

The solutions and answers that must be related to the problems and questions that are discussed here are dynamic and not permanent. They depend on the nature of the social decision-choice systems as established in different nations. The social outcomes, therefore, will vary over nations. The solutions and answers cannot be one shoe fits all due to the evolving nature of social dynamics. In the democratic collective decision-choice system, with a reasonable institutional configuration, the outcomes will vary as the collective preferences tend to shift or new generations tend to arise with differential collective preferences given non-deceptive social information structures. Different deceptive social information structures will generate different solutions and answers even in the same nation with the same collective preferences. The solutions and answers are generated in the political structure and given a temporary permanence by the legal structure until the decision-making core is reconstituted. This is due to the fact that the legal structure temporary locks a set of rights in the social set-up. This set of rights allows the social decision-making core to use the power of the state to protect its interest and enforce its will. The legal structure contains the legal ideology that is used to protect the decision-making power of the members of the decision-making core. Let us keep in mind that the government is abstractly neutral. The decision-making core is not. It is this non-neutrality of the social decision-making core that provide a justification to assess and revise the members through voting in a democratic collective decision-choice system and oppose to absolute monarchs.

It is the lack of sustainable permanence of these solutions and answers in the political structure that generates political and social instabilities in the democratic collective decision-choice systems which are established over the economic structure, where the decision-choice outcomes affect the political structure and then the legal structure. Every solution set is a temporary one waiting to be dislodged by new perceptions of reality and be replaced by new ones in the political decision-choice times for a new tomorrow under a different political risk. Thus, the social system is locked in a never-ending conflict behavior in the problem-solution duality. It is this conflict behavior in the problem-solution duality that propels the political economy in forward-backward transformations generating new qualitative characteristics on the basis of the conflict resolutions in the collective preferences over the public-private mix in the structure of the mixed political economy, whose environment for a democratic collective decision-choice system is supported by non-deceptive information structure. The possible distribution of public-private sector combinations and its relational structure to the conditions of distribution of freedoms in the individual-collective duality is shown in Figure 1.3.

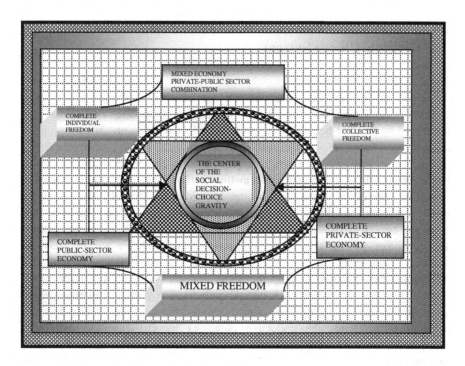

Fig. 1.3 A Cognitive Geometry of Possible Distributions of Public-Private Sector Combinations with Individual-Collective Freedom Duality in Defining the Environment for the Democratic Collective Decision-Choice System as well as Framing the Collective Decision-choice Problem

When the solutions are based on market imputations, the outcomes may produce distortions from the socially desired. The distortions of the market solutions are systemic and are shaped by the dynamics of income distribution, information structure, preferences of the community and the distribution of power in the political structure. The distortions may be amplified by introducing an element of deceptive information structure into the social information space. The corrections to the distortions are either carried on by the public sector through government assistance, in terms of subsidies and income support, or by the private sector through the combined institutions of philanthropic and non-profit organizations for the relief of market misery or by both the public and private sectors. These corrections are essential in creating and maintaining the stability of the political economy where the solution to the problem of provision of social goods and services is based on market imputations. The problem of these distortions and possible solutions must be related to the democratic collective decision-choice system that involves reconciling the conflicts of individual preferences without violence under the general social information structure which must be available to all.

1.4 Information Structure and the Democratic Collective Decision-Choice System

The selected regime of the private-public sector mix will always be determined by the social information structure and the set of individual preferences in a democratic collective decision-choice system. The individual preferences are influenced by the cost-benefit calculus of the collective decision-choice system in terms of the relative sizes of the public and private sectors with cost-benefit flows that are associated with the elements of the decision-choice set of private-public sector combinations. The general information on the cost-benefit characteristics is crucial for correct individual decision-choice actions and the collective outcomes by the application of democratic rules as established. Let us keep in mind that costs and benefits exist as duality in continuum for any decision-choice action. The outcomes of the democratic collective decision-choice system may be distorted completely to produce sub-optimal decision-choice actions by information manipulation holding the preference set constant. Any information manipulation has the power to change the individual cost-benefit imputations in favor of the unintended decision-choice actions. Furthermore, it also has the potential to influence the collective outcomes by changing the individual preferences that have been culturally shaped from the beginning in the social environment.

1.4.1 The Concepts of Defective Information and Deceptive Information Structures in Social Decision-Choice Systems

The social information structure is composed of *defective information structure* and *deceptive information structure* that we have already pointed out. They are distinguished by their roles and characteristics in decision-choice systems. The defective information structure is general to all decision-choice structures. It is made up of *fuzzy information* and *stochastic information*. The fuzzy information structure relates to language vagueness and representation, and ambiguities in transmission signals and thought. The fuzzy information is associated with quality of the general social information given the quantity of information. It generates *fuzzy uncertainty* and *fuzzy risk* that are associated with *possibilistic belief* in the decision-choice system. The stochastic information structure relates to information incompleteness, representation, transmission signals and thought in terms of volume. The stochastic information is associated with quantity of the general social information given the quality of the information. It generates *stochastic uncertainty* and *stochastic risk*, both of which are associated with probabilistic belief in the decision-choice system. The defective information structure is the sum of fuzzy and stochastic information characteristics. The total uncertainty in the decision-choice system is the sum of fuzzy and stochastic uncertainties. The total risk in the decision-choice system is the sum of fuzzy and stochastic risks where the total belief system is the sum of possibilistic and probabilistic beliefs. The *systemic risk* of the decision-choice system, therefore, is the total risk whose quantitative measure requires the use of methods of fuzzy and stochastic computing. As presented, the systemic risk of the decision-choice system is generated by the defective information structure within quality-quantity duality. This systemic risk is held to hold for all decision-choice systems. It may be amplified by the *deceptive information structure* in social collective decision-choice systems whether they are democratically or non-democratically organized.

The *deceptive information structure* is specifically associated with the collective decision-choice system of cognitive agents. It is made up of *disinformation* and *misinformation* substructures. It relates to language vagueness and representation, and ambiguities in transmission signals through information manipulations to change thoughts and direct preferences to the desires of the manipulators. Its effect is more pronounced in a democratic collective decision-choice system under the majoritarian principle of one person one vote relative to individual sovereignty. The disinformation process is a strategy to empty the mind of the decision-choice agents of what is known to be knowledge and to create *cognitive emptiness*. The misinformation process is a strategy to fill the cognitive emptiness of the decision-choice agents with information that may be made up of combined distorted signals to create a faked knowledge in support of particular decision-choice actions. Both disinformation and misinformation components of the deceptive information structure are associated with quality and quantity of the general social information given the defective information structure. The disinformation process reduces the space of the quantitative disposition of the stochastic information. This affects the quantity of knowledge possessed by the cognitive agent from available information. The result leads to the amplification of the stochastic uncertainty and the corresponding stochastic risk of the democratic collective decision-choice system.

In this respect, all social information classifications and restrictions belong to the area of disinformation. The misinformation reduces the quality of information thus enlarging the domain of fuzzy uncertainty and the corresponding fuzzy risk in the democratic collective decision-choice system. Similarly, propaganda of all kinds belongs to the domain of misinformation. The general purpose of the deceptive information structure is to create cognitive illusions in the minds of decision-choice agents through the development of information asymmetry in the democratic collective decision-choice system. The relationships among these information substructures are presented in Figure 1.4.

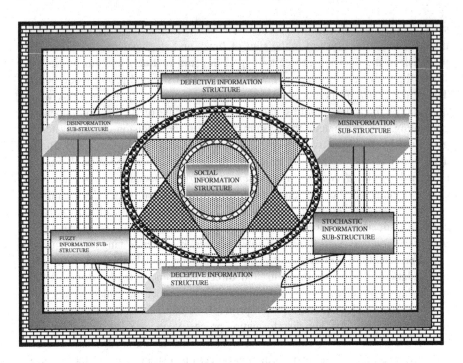

Fig. 1.4 A Cognitive Geometry of Social Information Structure in Defining the Environment for the Democratic Collective Decision-Choice System

The deceptive information structure thus amplifies the fuzzy and stochastic components of the social information, the corresponding total uncertainties and the systemic risk of the decision-choice system. The defective information structure creates penumbral regions of decision-choice actions with systemic risk. The deceptive information structure amplifies the penumbral regions of the decision-choice system and the associated systemic risk. In a collective decision-choice system driven from the political structure, the deceptive information structure is under the manipulation of those who have resources to engineer it. It is an indirect process to alter the defective information structure and temper with individual voting decision-choice actions. This is the area of information engineering that is under the propagandists and political strategists. Politicians and governments prefer to use the deceptive information structure to influence collective decision outcomes in the democratic decision-choice systems that works on majoritarian principles of voting to resolve the conflict in preferences in the collective decision space.

The interactive processes between the social decision space and the social information space generate information-decision-interactive processes which may be mapped onto the social risk space to generate a relational structure of information, decision and risk. The nature of the social risk system obtained from

the mapping of the social information system is shown in Figure 1.5. The structure
is such that the systemic risk is generated by the general system's uncertainty
which is produced by limited and fuzzy information sub-structures. The systemic
risk may also be amplified by the limitations of the information processing
capacities of individual decision-choice agents. The uncertainty-risk system that
has been presented here and elsewhere, suggests that any measure of systemic risk
on the basis of probabilistic logic will always under- estimate the true value of the
risk. Furthermore, our understanding and estimation of risk must proceed by
combining possibilistic reasoning and probabilistic reasoning [R7.14] [R7.15]
[R7.16]. The possibilistic reasoning allows us to deal with the qualitative character
of the information structure while the probabilistic reasoning allows us to deal
with the quantitative character of the information structure. The morphology is
seen by reference to the cognitive geometries of Figures 1.4 and 1.5. The defective
and deceptive information structures generate a complex fuzzy-stochastic
information structure in the democratic collective decision space which requires
an enhanced analytical tools of reasoning through a fuzzy-stochastic rationality
with its laws of thought and mathematics in the fuzzy-stochastic topological
space.

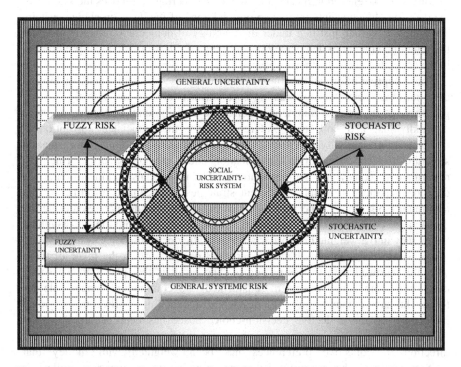

Fig. 1.5 A Cognitive Geometry of Social Uncertainty-Risk System in Democratic
Collective Decision-Choice Processes

1.4.2 Resources, Money and Deceptive Information Structure in the Democratic Decision-Choice System

The need for resources points to the notion that the deceptive information structure must be engineered and then disseminated through effective institutions and channels. It is the process of engineering and dissemination of the elements contained in the deceptive information structure and the costs that are associated with them that resources in terms of money and facilities find their way into the democratic collective decision-choice process with corruption. As it has been pointed out previously, the deceptive information structure is made up of two components. It is at the point of information dissemination, that mass media enters into the political market to generate the energy for carrying the elements of the deceptive information structure in order to transform the political market into an economic market to earn profits. Resources and monies are required for the transactions in the politico-economic market. These requirements favor the rich over the poor in the decision-making process. The effect of the process is to widen the penumbral region and increase the cloudy vile of the area of decision-choice elements, create additional conflicts in the debates to divide and conquer in order to construct consent through decision-choice engineering that moves the masses of the decision-choice agents into a sub-optimal zone of the decision-choice system. The outcome is then justified under the ideology of democracy rather than under the scientific content of democracy in resolving conflicts in individual preferences in the collective decision-choice space.

The deceptive information structure is not only used by individuals seeking to enter into the social decision-making core, but it is always and actively used by governments in matured democratic collective decision-choice systems to shape the individual and collective preference where the associated corruption is completely or partially legalized. The government's deceptive information structure involves information classification under national security principles, departmental secrecy and non-transparency. The individuals seeking to enter the decision making core employ different strategies of deceptive information that also require big money. The very nature of big resources and money entering the democratic collective decision-choice system alters it from an un-weighted democracy into a weighted majoritarian democratic decision-choice system, where vote castings carry weights between zero and one and are in the favor of those who are endowed with excess resources. The result is that places in the social decision-making core have been transformed into commodities under sale for those who have money to buy. The problem of reconciling conflicts in individual preferences in the collective decision-choice space has been accorded market solutions. In this way, democracy has been commoditized in such a way that the direction of collective preferences has been weighted with money that is controlled by the minority. By commoditizing the democratic collective decision-choice system, the principle of majority rule has also been transformed by the market imputation into the principle of minority rule by indirect purchasing and bundling of votes. In other words, the democracy is under the mockery of the three markets.

The deceptive information structure, its mass dissemination and its ability to manufacture consents and shape collective outcomes in the democratic collective decision-choice space must be seen in terms of social cost-benefit shifting that has important implications for income and resource distribution. The result of the process is that social costs are shifted to the least able and social benefits are shifted to the most able in terms of resource and income. A point of clarification is needed here regarding the information structures. The defective information structure is the central focus for information processing into a knowledge structure to support decision-choice actions in general decision-choice systems as has been discussed in [R15.13] [R15.14]. The information characteristics are inherent in all decision-choice systems and in all areas of knowledge production in support of decision-choice actions. In humanistic systems, deceptive information structure, intentionally engineered by cognitive agents, introduces complications in the social information structure by tempering with the fuzzy and stochastic components of the social information. The central focus is to alter the knowledge structure under the defective information structure in such a way that the deceptive information structure becomes the tool to influence the individual and collective decision-choice actions in the collective decision space.

The result of the use of deceptive information structure in the democratic collective decision choice system is to destroy the principle of the un-weighted majority rule in collective decision-choice systems based on non-deceptive information structures, and then replace it with a weighted majoritarian democracy where the weights are purely defined in monetary values. In this transformed weighted majoritarian democracy, the resource rich segment of the society dictates the terms of collective outcomes of the democratic choice system of one person one vote in the favor of the rich and powerful, and against the resource poor and weak segment of the society. Viewed in a more complex way, the *majoritarian democracy* has been destroyed and reshaped into a *minoritarian democracy* on the simple basis of market and money where every decision-choice action becomes commoditized into profit making. The outcomes of the money-weighted democratic collective decision-choice system is made to reflect the preferences of the minority, contrary to what is expected of an un-weighted majoritarian democratic collective decision-choice system, where, the process is the same, but the masses are deceived to give consent to decision-choice items that may have negative net cost-benefit values to them and even to the society, except the minority.

1.5 Costs and Benefits of Protection of Social Goals and Objectives for Rent-Seeking in the Democritic Collective Decision-Choice System

Having discussed the nature of social information and the role that deceptive information structure plays in the democratic collective decision-choice system, let us turn our attention to the discussion on the idea that the main instrument for the system of deceptions to be introduced in the collective decision-choice space is through information manipulation that will be available to the decision-choice agents. The deceptive information structure in the political economy is related to rent-seeking, rent-protection and rent-harvesting through the goal-objective game as it extends to the protection of the social goals and objectives. It is not enough in the social goal-objective game to win an inclusion of a group-interest specific goal or objective in the social goal-objective set. When the set is formed, each of the elements is under threat of policy exclusion by other political groups that seek to dislodge some objectives from the set and to replace them with their group-specific goals and objectives due to either resource constrained environments or ideological positions of other groups. All the elements in and outside the admissible social goal-objective set exist in a potential-actual duality in terms of policy practices and ideological shifts. As such, each element included in the social goal-objective set by whatever means must be actively protected by successful groups. Here, the *rent-creation game* with a *creation strategy* is transformed into a *rent-protection game* that requires a *rent-protection strategy*.

The rent-protection game has a duality process where coalitions of advocates of the private sector work diligently to unseat some social goals and privatize them for rent seeking, while at the same time, there are coalitions of public-sector interest advocates who work to protect the social goals and objectives in the public institutional domain from being dislodged by the private-sector interest advocates. Similarly, the coalitions of public-sector interest advocates works to include some specific social-interest goals and objectives into the social goal-objective set, while at the same time are working to widen the public-sector goals and objectives as well as to protect other social goals and objectives in the set from being dislodged by the opponents. At the decision level of the social organism, the private sector advocates and the public sector advocates exist in continual socioeconomic conflicts in the political and legal markets for shaping the direction of national history in terms of the production-distribution decision-making process in the economic market. The process is in a continuum whose influence becomes an important determinant of the resolutions of the temporary equilibria within the structure of the public-private sector duality of the social decision-choice process.

The private-sector and public-sector advocates are thus locked in a stochastically continual but point-to-point dynamic fuzzy game where each round presents a creation-protection duality in the formation and implementation of the admissible social goal-objective set. The reward of this game is the opportunity to create and protect rents in terms of possible net benefits which appear as fuzzy values. The creation and protection of the elements in the admissible social goal-objective set involve a construction-destruction process since the needed resources are limited for implementation of all goals and objectives that both private and public sector advocates may consider as socially desirable. The social desirability is seen in terms of a fuzzy sum of private-sector desirability and public-sector desirability. The activities of social objective creation and the protection of elements in the admissible social goal-objective set involve costs and benefits before even the social goals and objectives receive implementation.

The creation and protection of social goals and objectives, in general, present *rent-seeking possibility space* that defines an associated *fuzzy risk* of probable rent-seeking activity that is associated with *stochastic risk* in the democratic social decision-choice space. The implementation of any of the elements in the social goal-objective set reveals a *probability space* which defines the associated *stochastic risk*. The total risk in the democratic decision-choice system is that which is associated with the social system which we have referred to as the *systemic risk* that is composed of the sum of fuzzy risk and stochastic risk of the system's behavior due to the defective information-knowledge structure that generates possibilistic and probabilistic uncertainties in the system. This systemic risk may be substantially amplified by the deceptive information structure by enlarging the various risk components. The stability of the social system is specified within boundaries of the systemic risk and the tolerant level of the deceptive information structure [R7.14] [R7.15] [R15.13] [R15.18]. The enclosure of the boundaries of the systemic risk specifies the penumbral regions of social decision-choice actions. This enclosure belongs to the fuzzy-stochastic space of scientific work. The analytical tools for understanding the decision agents' behavior over this penumbral region must be abstracted from this space.

The costs of protecting the social goals and objectives are the social resources withdrawn from productive activities in order to protect the establishment structure of the social goal-objective set and the specific elements that provide possible opportunities for rent-seekers in terms of harvest. Since these resource costs are not directly related to socially productive activities or wealth-creation activities, they constitute social wastes in the organization of production and distribution. The social wastes appear as private benefits in terms of wealth transfers. If, however, it can be shown that such protective activities, like social infrastructures, indirectly enhance the productive activities of the wealth-creation process of the social organism, then we will have to consider the relative cost-benefit balances and the merits of the rent-seeking activities. Once again, it may be pointed out that these social wastes induced by the activities of creation and protection of social goals and objectives are not restricted to public monopolies and the social system of regulations. They are characteristics of collective decision making on the basis of *democratic* representation by a process where a decision-making core is established, where individual power to make social decisions is transferred to the members of the decision-making core, and where the members of the decision-making core have no immediate social accountability to the preferences of the majority of the voting public, except the interest of their own preferences and the preferences of the political salespersons. The interactive process of the rent seeking-game is shown in Figure 1.6 as a geometry of analytical thinking in dealing with the decision-choice problems in the political economy under fuzzy-stochastic information structures.

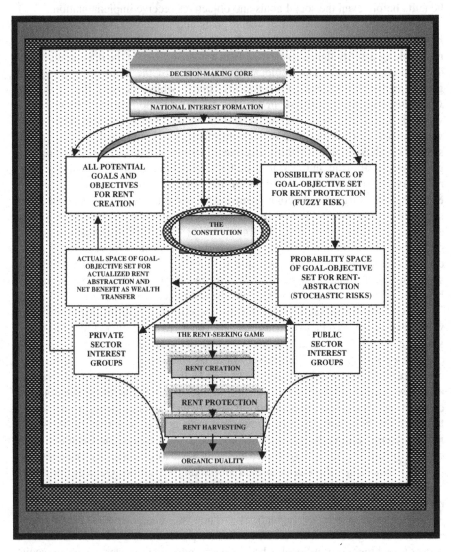

Fig. 1.6 The Geometry of Thinking of the Relational Structure of the Rent-Seeking Game and the Constitution under Fuzzy-stochastic Information Structure

It may be noted that the structure of the social costs includes the costs of lobbyists, and all kinds of consultants and lawyers in addition to the expenses in carrying on these activities of social waste. The social waste increases in volume as the political salespersons and rent-seekers increase in numbers and their activities acquire intense amplifications, where increasing resources are diverted from productive activities, and devoted by antagonistic coalitions of advocates on both private and public sectors, to buy influence from some of the members of the decision-making core. The rent-seeking and rent-protection game increases in complexity when both the private and public sectors are made up of a family of differential interest groups. This produces intra-private sector and intra-public sector competitions for rent creation and protection in favor of specific private interest groups (for example oil-production interest group, bio-fuel interest group, deregulation interest group, and others) or specific public interest groups (for example environmental interest group, smaller government interest group, poverty-reduction interest group or public-health maintenance interest group, defense expansion interest group and others). The decision-making core (the set of elected officials) also operates on the principles of voting under all kinds of accepted combinations of majority rule that must be indicated. Much of the society's resources are used in influencing the voting direction of the individual members in the decision-making core to cast their votes in the creation and protection of specific social goals and objectives that will serve group interests in terms of the possibilities of rent-seeking and harvesting. The whole conflicts in the collective decision-choice actions take place in the political sphere on the basis of political expediency and not on economic efficiency, even though the preferences are shaped by the distributions of cost and benefits in the economic sphere.

In an ideal situation, the casting of votes by the members of the decision-making core is to be guided by an indication and assessment of *public interest* without the influence of pressure groups. Under this ideal condition, the social cost of government would be the total compensation of the members of the decision-making core in addition to other expenses that are directly or indirectly connected to the social decision-making process. The total social opportunity cost is, thus, composed of internal and external opportunity costs. The total cost would be the sum of the costs of the best alternative productive and wealth-creating activity. This would be consistent with economic cost-benefit rationality in terms of cost-benefit calculations in the economic structure. But we do not live in ideal situations of politics, law and economics and we do not participate in the decision-choice processes in ideal conditions. Human actions, either individually or collectively viewed, are performed under conditions of a defective information-knowledge structure which is amplified by a deceptive information structure from the political and legal structures. The defective information structure, in addition to the deceptive information structure, affects the outcomes in the economic, political and legal structures as foundational blocks of social organisms. The presence of the defective information-knowledge structure, amplified by the deceptive information structure and operating with the fundamental ethical postulate of individualism, creates asymmetry of social power that affects the movement within the resource-expenditure duality and cost-benefit distributions in the social set up. All other resource expenditures by lobbyists and political salespersons over and above this total cost of social decision making are simply a social waste.

If the costs of influencing the creation and protection of goal-objective selection are wastes, then why does society permit them? The answer to this question may be found in the rent-seeking activities on the part of both the members of pressure groups as well as the members of the decision-making core, both of which extract rent and transfer wealth which benefit them as individuals or groups of individuals. These are undertaken at the expense of the general public in addition to the fact that the total cost is born by the member of the general public through a complicated system of taxation as the government revenue-generating machine. The political decision-choice actions have political cost-benefit configurations which lead to political cost-benefit rationality which guides the members of the decision-making core. In this political process, we have privatization of economic benefits and socialization of economic costs in the management of the social organism. The rent-seeking process adds to the voice of the neo-Keynesian macroeconomists that income distribution in the social production-consumption process is driven more so by social institutions, rather than by technical conditions of production as held by the neoclassical macroeconomists with their theory of production and income distribution.

It is precisely the privatization of rent-benefit possibilities and the socialization of rent-cost possibilities that prevent the American political decision makers from reforming the financing regime of the electoral process and politics as well as from holding on to the political duopoly. An immediate question arises. Will public financing of the electoral process eliminate the essential elements of private rent-seeking activities or reduce them and improve the democratic collective decision-choice process? Another way of stating the question is will public financing of the electoral process and an increase in the number of political parties will improve the political principal-agent problem of the social decision-choice process, and what kind of incentive structure must be introduced to ensure an efficient principal-agent relationship in the democratic social formation and institutional arrangements? We shall attend to these questions in the conclusion. Some other important questions are as follows: How do voters leverage their votes to affect the social income distribution through social projects? How do individual vote restrictions through all kinds of legal, political and economic conditionality in the democratic collective decision-choice system affect the social order and shape the national history?

Chapter 2
Rent-Seeking, Rent-Creation and Rent-Protection in Social-Goal Objective Formations

In Chapter 1, we discussed the hermeneutic aspects of rent-seeking in a democratic collective decision-choice system. The discussions allowed us to examine social information structure and the possible abstractions of the knowledge structure where the knowledge structure is seen as an input into the decision-choice actions in the political economy. The system was related to fuzzy or fuzzy-stochastic rationality, cost-benefit rationality, and expenditure-revenue duality. Let us now amplify the analytical works in the rent-seeking space.

2.1 Rent-Seeking, Expenditure-Revenue Duality and Influence-Buying Activities

Rent-seekers spend resources through pressure groups, lobbyists and political salespersons in the political structure to create a political space for rent-seeking activities. The reward from the expenditures by the pressure groups and lobbyists for pursuing activities of influencing social goal-objective settings by the decision-making core is the potential rent that can be actualized. The setting of the social goals and objectives and the supporting projects and programs are done through the decision activities of the social decision-making core of the political economy. These include defining the national interest, setting sub-social goals and objectives, creating projects, defining policy implementations and setting rules and regulations in the legal structure that affect social resource allocations, the markets, production-consumption decisions in the economic structure and the general and specific welfare of the political economy. In a democratic collective decision-choice system, the national interests, social goals and objectives are created and protected through various influences of interest groups in order to protect and actualize the potential rent inherent in each element of the goal-objective set resulting from influence-buying activities.

Changes in the laws of regulations to deregulation in the financial sector, the natural environment and resource exploitation are known examples of the results

K.K. Dompere, *Fuzziness, Democracy, Control and Collective Decision-Choice System,*
Studies in Systems, Decision and Control 5,
DOI: 10.1007/978-3-319-05329-5_2, © Springer International Publishing Switzerland 2014

of influence-buying activities in the goal-objective formation. Another example is the privatization of public productive assets for private profit making. The deregulation of the business of the financial sector opened up substantial areas of non-wealth-creation rent possibilities and social corruption under the legal structure, where the financial sector became transformed into a gambling scene of financial paper production in the United States of America and other Western Countries. This argument may be extended to markets for commodity futures that also acquire the conditions of gambling. The results were schemes to game the political economy for wealth transfers from more information-knowledge deficient users to less information-knowledge deficient users in the market of financial gambling that produced social waste without any real social production. The reduction in both extensive and intensive regulatory regime motivated the expansion of the deceptive information structure in support of the gaming activities to simply redistribute income, wealth and resources to the financial oligarchies. One of the results is the creation of social anxieties in various degrees and further reduction in the general productivity in the real sector.

The private benefits, which flow to interest groups in diverting productive resources to create the social waste in general, are the rents that emerge out of the game of goal-objective setting and the reduction of the regulatory regime. These private benefits have their cost supports that must be the responsibility of some members or all the members of the society through the collective revenue-cost system. From the viewpoint of the private-sector interest advocates, the struggles to include a particular goal and objective in the social objective set are driven by a simple private interest of rent seeking and protection to enhance profits, to reduce costs or to transfer wealth. The process requires the withdrawal of resources from socially productive activities to create a rent-benefit configuration and further withdrawal of resources to protect probable rents through the creation and protection of specific interest-group's goals and objectives, whose implementations require social resources, and whose benefits accrue to a specific group or groups but not necessarily to the public. In this process, there is asymmetry of dedication, strength and resources between private and public sector advocates in favor of the private sector advocates in capitalist societies.

The search for specific group-interest objectives for inclusion into the social goal-objective set is a search for rent-seeking of potential benefit that may or may not be activated. The creation of specific group-interest objectives in the social goal-objective set is the actualization of potential rent that must be translated into private benefits and directed to particular interest groups through rent harvesting. The resources devoted to the protection of specific elements in the social goal-objective set are the resources devoted to the protection of the actual rent that accrues to the private sector, and may fade into potential, if these group-interest specific elements are not protected for future harvest. The process of social goal-objective setting is dynamic and continually evolving. It induces cost-benefit dynamics through the social movements in the potential-actual duality of the elements of the set of social goals and objectives, where the actual may be dislodged into the potential and the potential may be actualized by the social game dynamics in the political economy. The process dynamically alters the social income distribution structure through the activities in the cost-benefit duality where the force of social conversion is the political moment. As it has been pointed out, the game results in a net loss to society and net benefit to specific private-sector groups. In other words, the total cost accruing to society is far greater than the total benefit accruing to the rent-seekers, and hence on the net, the rent-seeking activity is a net loss to the social system seen from the viewpoint of the decision-choice game of social goal-objective formation and implementation to bring the selected elements to fruition.

The driving force of the game is the potential rent as payoff to the private sector where rent actualization is sought to be translated into private-sector benefits. The maintenance of the game is through the dialectics of potential-actual duality under a continuum that encompasses each actualized element and each potential element under policy transformations. The logic of the ensuing analysis is that the discrete values generated by social decision-choice actions are temporary approximations of positions in the continuum which constitutes an enveloping of the temporary discrete values in the continuum of the actual-potential duality of the political economy. Every element in the constituted social goal-objective set has no claim to permanency even when it has been socially accepted and implemented by the social decision-making core. Each element exists in actual-potential duality. The elements may be voted out by the newly constituted social decision-making core as the social system moves through political time converting the actual into a potential and a potential into an actual by the changing elements in the legal structure and policy menu. Every element in the set of potential goals and objectives is a candidate for selection and actualization to dislodge an element or expand the elements in the goal-objective set by the decision-voting actions of the members of the social decision-making core. Every element in the set of the actual goals and objectives is a candidate for dislodging and potentialization into an abyss or for reducing the elements in the goal-objective set by the decision-voting actions of the members of the social decision-making core. The direct costs of searching for viably potential group-interest specific objectives, the actualization of the potential, the protection of the actualized and the potentialization of the actualized constitute the private sector cost of rent-seeking. The costs of organized or non-organized prevention against inclusion are the added private costs.

All these private costs are part of the real social costs of best alternative uses of the resources in the society for productive activities. In fact, an argument can be advanced in support of a position that a corrupt political system whether democratic or non-democratic deprives itself of an enlightened individual and collective liberty in favor of oppression of the weak by the powerful, just as inefficient uses of societal resources contribute to differential rates of economic growth and development over different nations, and over generations for the same nation. It is easy to abstract analytical support for this statement when power and information are seen as resources of production-consumption duality and management of a social organism as it relates to social income distribution and political stability. Since nothing is produced in the sense of real economic production, or, real wealth creation in the use of societal resources, then, these costs are *social wastes* whose benefits to the individuals or group of individuals are just transfers of either real income, or wealth from one sector to another, from one group to another, from an individual to other groups, or from the public sector to the private sector.

The total real benefit to society, if any, may be assessed in terms of weighted contribution to the welfare of the society in accordance with the moral code of conduct of fairness, justice and equality in the social formation and management. In fact, this moral code of conduct forms an integral part of the fundamental moral principles of either individualism, where the individual interest precedes the collective interest, or collectivism where the collective interest precedes the individual interest. The total social benefit contribution may be negative, zero or positive. What is clear is that resources have been wasted not to create wealth or improve organizational efficiency, but to create channels of society's wealth transfer that help to distribute and redistribute income and wealth among private groups and members as well as between private and public sectors. It is here that privatization of public productive assets, government outsourcing of public provision of goods and services, the tax-supported voucher system of public-sector provision of goods and services may be seen as wealth-income transfers from the public sector to the private sector for profit making.

2.2 Leveraging in a Democratic Collective Decision-Choice System and the Principle of Logical Continuum

The analytical process points to the notion that the democratic social decision-choice system is self-organizing and self-correcting which may assume the character of either self-destroying or self-improving. Every social system is under construction-destruction duality in continuum. The movements within the duality are carried out through the forces of social decision-choice actions and accelerator intensities. The intensities of these forces are shaped by the conflicts of individual and collective preferences within the individual-community duality. The position where the society will find itself, at any point of time, is a temporary equilibrium solution waiting to be moved by the next round of social decision-choice time. This position and possible changes depend on the voting preferences of the members of the social decision-making core and the public that creates them. Such voting-preferences may not be consistent with the public majority and hence not accountable to the basic principle of the public majority. It is here, then, that the ways and means to influence the members in the decision-making core arise. The instruments that are available to the voting public to influence the direction of individual preferences in the decision-making core without the use of violence are either vote leveraging, acquisition of influence in the decision-making core or both. The acquisition of influence is associated with influence-buying activities which are limited to the resource-rice minority. What is available to the voting public is vote leveraging for each round of the political cycle where the voting public becomes the agent and the actual and potential members of the decision-making core become the principal in the principal-agent duality.

Political leveraging is an essential tool for affecting the direction of change of the elements in the social goal-objective set in the democratic collective decision-choice system, and hence used to affect the nature of the social cost-benefit distribution without violence. There are two important types of political leveraging in the collective decision-choice system. One is *voting leveraging* and the other is *influence leveraging*. Vote leveraging first requires seeking voters with the same interest on a specific element or elements in the social goal-objective set. Then, the next step requires creating a *block voting* to either persuade or threaten the members of the social decision-making core to act in favor or against an item to be voted on. The democratic collective decision-choice process, given a non-deceptive information structure, defines the initial conditions for the practice of a democratic collective decision-choice system under the fundamental ethical postulate of individualism with a set of practicing conditions. These conditions, even though necessary, are not sufficient in the sense that they do not establish successful practice of democracy in the space of the political economy. The relationships among the people, government, social decision-making core, the governance and the democratic collective decision-choice system have been explained in [R15.18].

When the democratic collective decision-choice system has been successfully used to resolve the collective conflict in individual preferences to establish the social decision-making core as its initial condition, a new collective conflict arises in the majority-minority duality. This collective conflict in the majority-minority duality under a transformation continuum where a role reversal is possible in the sense that the minority may be transformed into the majority and the majority may be transformed into the minority by the decision-choice process. The conflict is the turned into a game to influence and shape the direction of the preferences of the members of the social decision-making core who are in favor of either the majority or the minority. It must be kept in mind that the members in the decision-making core have their own preferences that may be compatible or incompatible with the majority or the minority or neither the majority or the minority of the voting public. To influence and shape the directions of the preferences of the members in the social decision-making core, some leveraging is needed by the voters that act as the agent during the period to constitute the social decision-making core, who substantially is now transformed into the principal by the rules of the democratic collective decision-choice system as provided by the ruling constitution.

It may be understood that the social decision-choice actions of the members of the social decision-making core may go against the will of the majority as revealed by their preferences and in favor of the will of the minority, depending

on the instruments of leveraging in the decision-choice actions of the members of
the social decision-making core. After the collective decision to constitute the
social decision-making core, the next step in the democratic decision-choice
system in relation to the elements in the goal-objective set and the national interest
is to activate the instruments of leveraging, in order to pressure the members in the
social decision-making core to act in a particular direction in selecting the
elements of the goal objective set. It is at this juncture that an organized resource
enters. The resource enters to create block votes in terms of positive actions to
influence the preferences and decision-choice actions of the members of the
decision-making core. The positive actions come into the political scene as
demonstrations, civil disobediences and vote withdrawal threats for the next
voting cycle and against some members for their voting actions in the social
decision-making core.

Similarly, the resource for leveraging may enter as vote-buying from the
members of the electorate and may also influence buying from the members of the
social decision-making core in more or less legal or illegal political markets. It is
here that money enters to shape and infest the democratic decision-choice process
in the collective decision-choice space with corruption and injustice. The buying
of votes and influence is directly linked to rent-seeking in the political economy to
which we turn our attention, to argue that rent seeking is part of the capitalist spirit
that motivates all kinds of arguments for small government and privatization. The
whole influence buying-selling activity is to create an economic and financial
space for profit taking and enhancement. In this respect, the organization of
provision of goods and services in support of individual and collective existence
for collective benefit has been transformed, under the market system, into an
organization of profit game with losers and winners. The organization for the
provision of increasing collective wealth accumulation for the benefit of all is also
transformed into the organization of increasing individual wealth accumulation for
the benefit of few.

The argument for privatization, small government and out-sourcing is an
ideological support for the creation of this economic space. The alternative
argument in support of public-sector size is also an ideological support. Both of
these ideologies can be illustrated with the diagram of public-private sector
efficiency frontier as in Figure 2.1 and the bimodal distribution of ideological
preferences over the private-public sector proportions in the political economy.
The 45°–line defines an ideological balance. The public-private sector frontier
defines the set of all possible proportionality combinations that may be used to
develop institutions for the provision of social goods and services. To the left of
the ideological balance is an increasing public sector size and to the right of the
ideological balance is an increasing private sector size. Nationalization rotates the
line of institutional creation and support of provision of social goods and services
to the left of the balance. Similarly, privatization rotates the line of institutional
creation and support of provision of social goods and services to the left of the
balance.

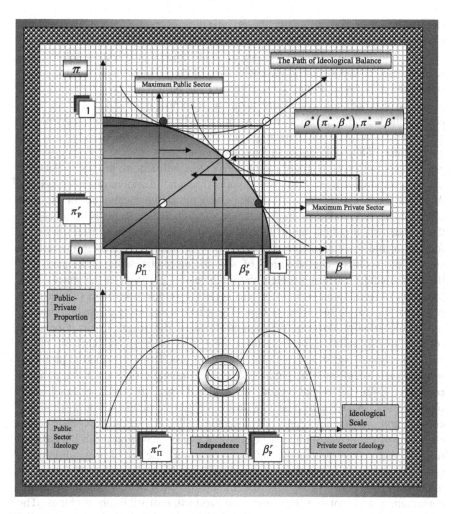

Fig. 2.1 The Private-Public Sector Efficiency Frontier, Independence and Ideologues in the Political Economy where π = public-sector proportion, β = private-sector proportion, $\pi_{\mathbb{P}}^r$ = public-sector reservation proportion by the private sector, β_{Π}^r = private-sector reservation proportion by the public sector, $\pi_{\Pi}^r - \beta_{\mathbb{P}}^r$ = ideological distance and zone of negotiation

2.3 Rent-Seeking and the Argument for Smaller Government and Increased Privatization

It may be pointed out at this point, that the use of wastefulness of the rent-seeking process composed of creation, protection and harvesting to argue, as it is done by a number of rent-seeking theorists on government and the political economy [R13]

[R13.1][R13.5][R13.6] to justify the privatization of public sector activities and the down-sizing of government no matter how the size of government is measured in order to reduce social waste, is more of an ideological justification rather than a scientific one in support of an increasing size of the private sector and decreasing size of the public sector. The same argument can be levied against governmental activities of democratic social organization (refer to Figure 2.1). Let us try to abstract the logical sequence that the argument is built on, in support of deepening privatization as well as widening the size and the size reduction of the government. This will allow us to understand and abstract the cost-benefit possibilities of the social goals and objectives that will require program selections by the method of relative costs and benefits through a democratic-choice system. As it has been explained in [R15.18], given the defined elements in the goal objective set, the implementation of the decision-choice elements is designated to the private or the public sector on the basis of which either new institutions are formed, or old institutions are strengthened or dismantled to change the relative size of the sectors. The selection of the private-public sector combination is the institutional formation problem that involves the partition of the space of the provision of goods and services to the society by allocating them to either the private or the public sector or by both. The problem is simply the public-private-sector allocation-production- decision problem of the social set-up. Given the solution to this public-private sector combination problem the analysis of the political economy must deal with the allocation-production-distribution problems of the private sector and the public sector. The organic efficiency in the political economy, given the social vision, the national interests and the social goal-objective set, must be seen at two levels of private-public-sector institutional combination efficiency and intra-sector allocation-production-distribution efficiency of the private-public-sector institutional combination given the national resource endowment to execute the social program. The social program is made up of 1) setting the national interest and defining the social vision, 2) setting up the social goal-objective set, 3) solving the private-public-sector institutional combination and solving the allocation-production-distribution problem. The solution to the social program in relation to either both private-sector expansion and deepening or both the public sector expansion and deepening is ideologically and value-dependent. Let us now examine the logical steps offered by the private-sector advocates for private-sector deepening and expansion for increasing its size.

2.3.1 Logical Steps of the Argument for Smaller Government and Non-market Interference

The logical steps of the argument in support of government-down sizing, reduction in the relative size of the public sector to private sector through privatization and outsourcing of public-sector provision of goods and services run as follows:

1. The greatest barriers to private entry to the production system and increasing socioeconomic activities of the society is the government and government's created regulations and rules.
2. These government created barriers generate, define and maintain actual and potential rent that individuals and groups of individuals may seek (that is, government regulations produce rent-seeking possibilities).
3. There are two types of rents that individuals or groups in the private sector may pursue. They are: a) wealth-creating rent-seeking that is socially productive, and b) wealth-transferring rent-seeking that is socially unproductive and wasteful.
4. The government is the major and most important source of wealth-transfer rents as well as promoter of rent-transferring activities and hence creates social waste by its very existence.
5. The "bigger" the government, the greater is the social waste due to wealth-transferring rent activities.
6. The impersonal market promotes wealth-creating rent activities and hence socially productive, efficient and desirable.
7. The government's interference in this impersonal market introduces government wasteful activities into the market, and hence the less interference by the government in the market system, the greater is the wealth-creating rent activities.

Any of these steps of the argument in support of government down-sizing and privatization of public-sector productive activities may be expanded into many different dimensions. All of the arguments are based on private-sector ideology in the public-private sector duality, social production of goods and services and incentive structure on the simple principle of individual cost-benefit rationality, with little reference to the individual-community cost-benefit rationality as has been presented in cognitive geometry in Figure 2.1. A similar set of steps of arguments in support of public sector expansion and increased relative size can be structured for a comparative analysis. The essential difference rests on the incentive structure of profit-nonprofit duality within the private-public sector duality in continuum for the distribution of income and wealth among the members of the social set-up.

On the basis of similar reflections on the market as the most efficient solver of socioeconomic problems, the Research and Policy Committee of the Committee for Economic Development (CED) composed merely by business elite had the following to say in the Redefining Government's Role in the Market System, published in 1979:

1. *The expansion of government's role in the markets has in many cases impaired the performance of the U.S. economy.*
2. *The trend toward accelerating inflation, lower productivity, and sluggish economic growth has been exacerbated by the expansion of government*

programs and by regulatory policies that reduce the productivity of the economy.

3. *The growth of some public expenditure programs and some indirect incentive programs has made it difficult to reduce large federal budget deficits. Regulations have increased the costs of production, reduced productivity, discouraged capital investment, and raised prices, often without providing adequate compensating benefits.*

4. *Government policy makers have often failed to recognize that the market system is an effective way of achieving policy goals and that the output of the market system must be permitted to expand in order to provide benefits to the public—consumers, workers, and shareholders.*

5. *Improved economic performance and the achievement of social goals require that government involvement be reduced and that ineffective and inconsistent government programs be eliminated.*

6. *When government involvement is justified, markets should be used whenever possible to implement policies. Restricting the operation of the market system or failing to encourage competitive markets is often the most inefficient way of achieving desirable economic and social goals simultaneously.*

7. *Weaknesses in the current policy-making process within the political system have initiated and continued to support the trend toward ineffective government interference with markets.*

8. *The goal of slowing the growth of aggregate expenditure should be supported. This type of constraint on the political decision-making process will not by itself solve the problem, but it will help.*

9. *The decision-making process for public policy must provide policy makers with the information to reduce government's role and to modify counterproductive expenditure programs and regulations systematically.*

These steps when accepted, in addition to the utility of increasing wealth, lead to two sets of prescriptive rules both of which are for government management since the market is assumed to be impersonal and efficient by its operative mechanism. One of the prescriptions is in terms of the structure and form of the government itself. The other prescription is in the form of the behavior of the government toward the market structure where the market imputations are taken to be neutral, efficient and unbiased. The general thrust of the argument is that economic waste flows from the public sector and may be eliminated by reducing its size and by implication the size of the government. The implied logical derivative is that poverty is not the result of inefficient functioning of the markets but inability of the poor to take advantage of the market. The implied set of the statements in the argument is a libertarian approach in resolving conflicts of individual preferences in the individual-community duality through either the minimization of, or reduction of the importance of the role of the government and hence the public sector in the collective decision-choice space.

The emphasis is placed on the power and efficiency of the market to resolve conflicts in the individual preferences and to blame the government if the market solutions are socially unacceptable. The proponents of this line of thought fail to see that collective actions through unity expands the set of decision-choice alternatives over which the individual or the collective may operate through the public sector. The individual decision-choice sovereignty and the collective decision-choice sovereignty will always be in conflict since the individual sovereignty is a constraint on the collective sovereignty, and the collective sovereignty is also a constraint on the individual sovereignty in the social decision-choice space. It is precisely this conflict in the individual-collective duality that provides the social system with the energy and forces for transformational dynamics with qualitative and quantitative dispositions. The nature of the conflict and engendered transformational dynamics is illustrated with the diagram of Figure 2.2.

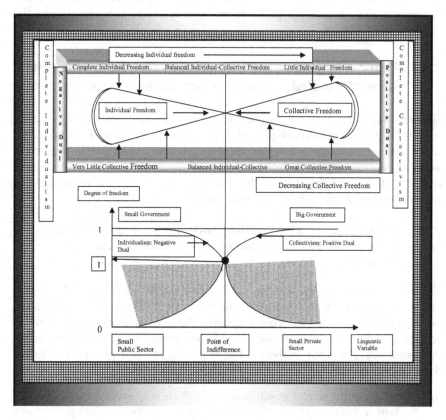

Fig. 2.2 An Epistemic geometry of Private-Public-Sector Duality, Individual-Collective-Freedom Duality and Small-Big Government Duality in Continuum and in Relation to the Fuzzy Laws of Thought and Approximate Reasoning

The resolutions of the conflicts in the individual-collective-freedom duality solution to the public-private sector combination and the solution to the problem of big-small government are combined in a geometric illustration under the conditions and logic of fuzzy rationality. The involved problem may be analytically formulated as fuzzy optimization problems. The problems may be formulated in different ways where big-government, small government and individual and collective freedoms are seen as fuzzy sets with appropriate membership characteristic functions. The size of the private sector may be taken as the objective and the size of the government as a constraint for the fuzzy optimization. Similarly, the degree of individual freedom may be taken as the objective and the degree of the collective freedom is taken as a constraint for fuzzy optimization. The objective-constraint system may be reversed as the social preferences shift. Figure 2.2 gives an example of the logical process and the corresponding solution.

2.3.2 Rent-Seeking and the Prescription by the Market-Interest Advocates to Reduce Social Waste due to Government Size

Given the logical steps of the argument by the private-sector interest advocates in support of government-downsizing, reduction in the relative size of the public sector and increased private sector through privatization of public productive assets and outsourcing of the public provision of goods and services, the following recommendations are offered by the rent-seeking theorists on government to reduce social waste through public-sector reduction.

2.3.2.1 Prescriptions to the Social Decision-Making Core for the Structure and Form of the Government by the Private-Sector Interest Advocates

The private-sector-interest advocates carry with them the ideology of the market as a social instrument to solve the problems in the conflict space of collective preference in social decision-choice system. The goal is to expand the private sector and reduce the public sector by using the market wherever possible in the three structures for profit motives.   The following are their prescriptions to the decision-making core:

a) Reduction of the size of government is a reduction of social waste caused by the rent-seeking activities involving only wealth-transfers. The government must reduce its size since large size creates large social waste due to the government's potential in creating a social environment for rent-seeking activities that are non-wealth creating.

b) Governmental activities and projects in the non-military sector must be privatized or out-sourced to reduce the "size" of government, and even in the military sector privatization must take place whenever it is useful through contacts and outsourcing to the private sector.

2.3.2.2 Prescriptions to the Government Regarding Its Behavior toward the Market System

a) The government must reduce its interference in the market system, by deregulation, since deregulation increases the potential wealth-creating rent that invites innovators and imitators, in the sense of *Schumpeterian dynamics* from one *Shumpeterian equilibrium* [R4.42][R4.57][R4.99][R4.100][R4.101] [R4.103] to another through the internal dynamics of the market system and private sector behavior, on the basis of self-interest and the fundamental ethical postulate of individual freedom. This will increase efficiency of resource usage, reduce social waste and the social cost of government leading to an increasing social benefit and improved overall social welfare of the society.

b) The government should follow the policy of hands-off productive activities and the creation of social goods and leave these activities and the production of social goods and services to the goodwill and efficiency of the private sector through the use of benefit incentives and cost disincentives.

c) The government should promote private-public partnership by establishing institutions that outsource the government's provision of goods and services from the public sector to the private sector. Some of the methods for accomplishing this include the use of government contracting, denationalization, positive tax incentives, grants and vouchers and other incentives that enhance the private-sector activities.

The logical derivation from these steps and prescriptions leads to the conclusion that the social goals and objectives of the nation must be set and implemented to favor private-sector benefits at the expense of public-sector benefits. The process in its economic computing logic is politically transferring potential public-sector benefits to the private sector and the private sector costs to the public-sector in the provision of the national goods and services in fulfilling the elements in the goal- objective set. Alternatively viewed, the members of the decision-making core must socialize private-sector costs and privatize the public-sector benefits to the extent to which social stability is maintained. The public sector generates waste but not wealth. When the private-sector benefits increase, the social system will benefit indirectly through a trickle-down process. The second order derivation leads to the sub-conclusion that social goals and objectives must be set and implemented by the government to socialize costs of the private sector inefficient activities that result in losses due to mismanagement and bad private-sector decisions.

Profits from both public and private sectors that result from rent-seeking activities which are rightly or wrongly considered as wealth-producing must be privatized. The cost of social infrastructure in support of the private sector rent-seeking activities must be socialized. The costs of public-sector investments to widen the space of opportunity must be socialized, while the facilities and their productive enterprises and services must be outsourced through a system of public vouchers to the private sector for profit making. Thus, the structure of the costs and benefits of the organization of society for the production and distribution are directly tied to the social goals and objectives and their implementations. According to the private-sector ideologues, the social goals and objectives must be set to create an appropriate social incentive system in the market for private sector rent-seeking that is wealth-producing. The distribution of such costs and benefits are completely directed by the structure of social goals and objectives that are created and implemented by the social decision-making core in the democratic decision-choice system and its corresponding institutional support. Thus, by associating the social goals and objectives to private-sector incentive structures, the government will promote wealth-producing rent-seeking and efficient resource usage. This is one of the important reasons in support of the position of the Neo-Keynesian economists that cost-benefit distribution or income distribution in the society is institutionally determined[R4.33] [R4.48] [R4.78][R15.41][R15.42] as opposed to the position of the neoclassical economists where the distribution of costs and benefits of social production is technologically determined [R4.41] [R4.18] [R15.2] [R15.31]. The cost-benefit distribution, just as the power distribution in the collective decision-choice space or the distribution of socio-political freedoms, cannot be institutionally free. As it has been argued in [R7.14] [R15.33], every decision-choice action of social significance takes place within a given culture that contains its moral values and institutions. It is these values and institutions that give meaning to the distributions of income, wealth, opportunities, freedom, and poverty within any individual-community relativity in the social decision-choice space.

By critically examining costs and benefits associated with the creation, preservation and maintenance of objectives within the formation of social goals and objectives, given the national interest, it seems clear to indicate that rent-seeking activities are part of all government operations in all the fields of economics, politics and law. If the *size* of the government is defined in terms of *expenditure size,* then a reduction of social programs, increase in military programs and privatization of the production of social goods are simply resource transfers and alternative distribution of potential rent and national wealth. Such transfers have social costs and benefits that are tied to the social goals and objectives as created and implemented by the social decision-making core. Here, it may be argued that the greatest potential rent does not reside with social programs and regulatory activities of the government, as we have been made to believe by some economic, legal and political theories on government.

The whole government apparatus is nothing but institutions of control and regulation conceived to be for the protection and defense of the social vision, public interests and welfare. Furthermore, the greatest potential rent resides in the government created special privileges for the producers in the defense and special industries. The defense-sector activities are vehicles for waste-creation and wealth transfers from the tax-paying public to private-sector profit seekers. A case can be made that the transfers of production of goods and services from the public sector to the private sector generate abuses and scandals that lead to increased systemic risk and government bailouts with further socialization of costs with increasing abuse of collective freedom. In this connection, certain questions arise. How does the private sector make profits from the transfers of the provision of services and output from the public sector? How does the private sector get paid for the provision of public-sector outputs and services? Are these profits not from the social revenue of taxes? Let us relate these questions to the claimed objectives of privatization.

2.3.2.3 Objectives of Privatization for Smaller Government by the Private-Sector Advocates

The whole idea of privatization is a set of decision-choice actions to accomplish a set of objectives. Some of the important objectives of privatization are stated below.

1. The improvement of economic performance of productive capital;
2. The de-politicization of economic decisions;
3. The generation of public-revenues through sales of public enterprises for private-sector production;
4. The reductions of taxes, government borrowing needs and public provisions of social goods and services;
5. The reduction of the power of public–sector unions and their impact on the politico-economic process; and
6. The promotion of capitalism through increasing private ownership of productive capital.

Generally, privatization decision-choice actions are strategies to shrink and control the government size and its growth in the sense that if private-sector provision of goods and service can be accomplished with fewer resources than the public-sector provision, then the private-sector should be given the opportunity for the needed provision. It may be observed that political, economic and legal decisions are not mutually exclusive. The concept of privatization is fully involved with unity of economic, political and legal decisions. It also relates and affects the income-wealth distribution among the members of the social setup. None of the above precepts can be accomplished in only one of the three structures of economics, politics and law. Every decision, either social or private involves cost-benefit balances which are mapped into the space of preference ordering and harvested in the economic structure.

Privatization as a vehicle to improve economic performance of public-sector productive capital is based on unproven claims that private ownership is more productive than public ownership of productive assets. This is more of an ideological claim than a scientifically proven statement. The efficiency of private ownership of productive assets in general is more or less explainable from institutions of income distribution where part of labor's productivity is appropriated for capital in terms of profit. The whole notion and concept of depreciation of factors in use should be reexamined. What is the justified reason to allow depreciation expenses to be applied to capital but not to labor? The claim that privatization leads to de-politicization of economic decisions is simply false since privatization is a political decision and no social decision-choice action can claim independence from the politico-legal process and cost-benefit rationality. The objective of generating public revenue through sales of public productive assets for private-sector production is related to the objective of the reduction of taxes, government borrowing needs and public provision of social goods and services. While it is true that the public-sector provision of goods and services will be reduced, the objective of reduction in taxes and the government borrowing needs may not be fulfilled especially when there is a voucher-subsidy system supporting the private-sector production of the goods and services.

The objective of the reduction of the power of public–sector unions and their impact on the politico-economic process is extremely important in power distribution within the democratic collective decision-choice system. The idea is to weaken the labor participation and enhance the private capital-owning class through increasing the power of capital's effect in the democratic collective choice system. The increased power of capital and the reduction of the public sector lead to the promotion of capitalism through increasing private ownership of productive capital and the reduction in collective ownership of productive capital in the social set up. One important thing about the whole size of the government is the failure to acknowledge the public-private sector duality with a continuum. The public-sector investments go to enhance the private-sector productivity depending on the distribution of the investment over public productive and non-productive activities. Private-sector expansion through out-sourcing and voucher-subsidies of public-sector production of goods and services for private profit-making, and paid through the people's taxes is simply another form of income-wealth transfers at the expense of the public sector. A question arises in the private-public sector rearrangement. Do an increasing privatization and a decreasing public sector enhance and increase the social welfare of the social set-up where such social welfare is income-wealth-distribution dependent? What is the relationship between the set of private-public-sector combinations and the social welfare of the society? How are the various private-public sector combinations related to the social vision, the national interest and the social goal-objective set? How is the social welfare related to various levels of wealth-income distribution? How is the private-sector efficiency enhanced by the public-sector provision of goods and services such as social infrastructure of various forms? The answers to these questions are very important in any claim of either private or public sector superiority in efficient provision of goods and services.

2.4   The Nature of Rent-Seeking in the Capitalist Socioeconomic Organism: The Real and Financial Sectors

Rent-seeking activities in relation to the elements in the goal-objective set must be seen in terms of their effects in financial and commodity sectors (the commodity sector includes military production). One thing is historically revealed: an unregulated market system has a tendency to produce extreme outcomes where only the strong and powerful will always benefit. Social resources are then withdrawn to create strength and power in the social set up by whatever means that the system will allow. Regulation is a necessary evil for the economic, political and legal structures in the sense that it restricts individual and collective freedoms in the collective decision-choice space where the individual-collective freedoms exist as mutual constraints on individual-collective preferences. The existence of the legal structure projects a system of regulations of all social behaviors in all the sectors. The question that arises in the political economy is not whether regulation is good or bad but what social and private decision-choice behavior must be regulated?

Regulation, just as the government, is good-evil duality that is translated to benefit-cost duality in a continuum. The role of the government as an institution is the complete management of the society through regulations and enforcements of the democratic decision-choice system for both the individual and the collective. Both the private and public participations in the social setup are constrained by political institutions through the legal institutional structure. If you accept the economist's model of a perfectly competitive market system and project it to the political and legal markets then, the rent-seeking activities are completely replaced by the profit-seeking activities based on unregulated self-interests on the fundamental principle of unregulated individual and private freedom. Let us keep in mind the concepts of duality and continuum where no regulation is regulation in duality. To see how rent-seeking emerges out of the social system organized on the fundamental principle of individual freedom, interest and markets for profit motives, let us present the basic characteristics of such a social organism. In this process, we must keep in mind that the government is simply an institutional facility and a holding place. The social decision-making core is the occupier and user of the institutional facility to accomplish certain social goals as well as facilitate the collective decision-choice process. The actions and the process of the socio-political decisions constitute what is called the governance which is distinguished from the government. The implementation of the elements in the social goal-objective set may be undertaken solely in the private sector or solely in the public sector or in a selected public-private sector combination in order to accomplish the national interest and the social vision. Let us examine the solution provided by the institutions of a perfectly capitalist political economy as a framework to examine a political economy with various private-public sector combinations.

2.4.1 The Basic Characteristics of a Perfect Capitalist Socioeconomic Organization

The basic characteristics of a democratic social decision-choice system in perfectly competitive capitalism begin with the building blocks with the following structure:

1) An economic organization based on democratic individual freedom to choose in the *economic market place* to satisfy its private interests;
2) A political organization based on democratic individualism where there is the individual freedom to operate in the *political marketplace* to serve individual interests; and
3) A legal organization based on democratic individualism where there is individual freedom to operate in a *legal market place* to serve individual interests.

For these markets to project an organizational unity, they must have the characteristics of consumers and producers in a duality setting, where the markets function as institutional organizations based on individual willingness to choose, conditional on his or her resource endowments without force. The interactive nature of producers, consumers and markets is defined by three sub-structures of decision, information and motivation. The outputs of these markets differ from one another but are inter-supportive for the social outcomes.

The decision-choice substructure in this perfectly competitive behavior in the production-consumption duality specifies the conditions that the legitimate power of decision-making in each of these markets is vested in the individual or group of individuals, where the power to choose is an individual right based on the fundamental moral postulate of individual self interest without regard to the collective. This is the *postulate of individual freedom* to choose in a perfectly competitive market system without government. Here, the aggregate decision-choice action is an unexpected outcome of interactions of uncoordinated individual decisions and choices. There are no rules and regulations except those of self-interest. The aggregate outcome is the result of interplays of relative powers of vested interest groups. The social system and its decision-choice structure are automatically regulated and unconsciously managed by the interplay of antagonistic individual self interests. Technically, the security system is organized on the basis of individual interest and market imputations.

The decision-making process must be supported by an information sub-structure. The characteristics of the information sub-structure are summarized by prices and income or resources. These markets have sets of prices or exchange ratios that provide a summary of past and present information sets about the items of exchange. The information set may be expanded to include current, forward and future prices and commodities, viewed in a broad context of exchange. The information sets are generated through evaluative interactions of individual agents in the markets. The individuals are responsible in obtaining and processing the information into individual knowledge structures. Given the information sub-structure, the decision-choice process is supported by a motivation substructure which is made up of selfish interests of individuals or groups of individuals. The selfish interests may be of material or non-material types. The aggregate interest that comes to rule in the political economy is the unexpected result of competitive interactions of individual self-interests that define different preferences in the provision of goods and services in the political economy.

Each of the markets is, here, viewed as institutional arrangements through which the conflicts of individual preferences over choices of alternative commodity-service bundles are resolved by the actions of self-interests under a given information regime. The individual and collective outcomes and the optimal decision-choice processes are extremely sensitive to the nature of the information regime. In this respect, economic, political and legal markets are said to exist if the following conditions are met by each one of them.

a) There are definable services or commodities that are exchangeable;
b) There are social agents who interact in the political economy for the purpose of acquiring or disposing of commodity or services in such a way that the suppliers, demanders, producer and consumers are defined;
c) The interactions are in the form of voluntary exchange of commodities or services for other commodities or services (or money or acceptable medium of exchange) where such an exchange is based on the fundamental ethical postulate of the individual freedom to choose on the basis of self interest but not on the basis of collective or social interest;
d) Communications among the interested agents take place in terms of prices and quantities that must be defined in a specific sense for any given endowment of the individual and the collective; and
e) Competition exists between the buyers and sellers, or the producers and consumers and among themselves within the groups.

The establishment of the goal-objective set that defines the environment of rent-seeking is first examined around the economist's perfect competitive markets that follow the following characteristics:

1) There are two sets of producers and consumers in each market who meet the existence of consumer-producer duality;

2) Each market is made up of a large number of sellers and buyers (the postulate of atomicity in decision-choice action);

3) Individual decision-choice units act "as if" they have no control over prices or the information regime and hence take prices or the information as given except when they act in groups under given income limitations (the postulate of price taking or information independence);

4) There is complete and perfect information in all three markets of economics, politics and law (this is the postulate of non-deceptive information structure). The postulate of perfect information structure translates into:

A) Each buyer has an exact and full knowledge of all information relevant to his or her buying activity or exchange transaction in relevant markets. For example, in the political market the politicians know the type of influence that they are directly or indirectly selling and the buyers know the type of influence that they are directly or indirectly buying;

B) Each seller has an exact and full knowledge on all information relevant to his or her selling decisions in all relevant markets;

C) The exact and full information implies that in the markets, all decision-choice agents have:

a) full information of the choice set and limitations on it; b) full and precise information of technical capabilities of each object in the choice set to satisfy a want or goal-objective element; c) full information of exact unit cost of each choice object; and d) full information to the fact that individual independent action will not change the prevailing market unit cost and the market unit benefit.

5) Each decision-choice agent has a regular preference function and an access to the factors that shape the structure of the socioeconomic organization, its institutions and markets.

When all these characteristics are applied to the economic structure of a society without the political and legal structures, the outcome is a *chaotic decision-choice system*. In such an uncontrolled market system, the only rule of behavior is the pursuance of individual self-interest, where survival of the fittest is its natural law in the collective civility that operates closely to the evil side of good-evil duality, but not on the basis of social laws that move decision-choice agents to operate closely to the good side of the good-evil duality in human existence. The individual interests are not constrained by the collective interests. To eliminate the chaotic character of the uncontrolled economic structure, an organization of the people into civil social groups is required. Such an organization resides in the political structure as an institution of government. The establishment of the political structure and its institution of government require managers and political administrators to operate the governmental machinery. It further requires rules and laws for organizing and managing the social set up. These rules and regulations are created in the political structure by the social decision-making core and codified into laws in the legal structure that comes to regulate the boundaries of individual and collective behaviors in the pursuance of self interests in all three structures for social stability, internal peace and collaborative work. The managers and administrators, who together constitute the social decision-making core, are the makers of the rules and regulations in the political structure. The mode of the selection of the members into the social decision-making core, to occupy and use the institutions of government, varies from one political economy to another and from one generation to another, even though the institution of government remains the same in concept and operation. The management and the administration of the political economy together constitute the activities of governance which will vary from one social decision-making core to another for the same political economy and the same institution of government.

The establishment of rules, regulations and laws, defines the parameters of self-interest behavior in relation to the collective interest, as well as creates environments of rent-seeking and corruption in the three structures depending on the administrative and managerial style of the members of the social decision-making core. These rules, regulations and laws affect the lives of all individuals in the society and their allowable spaces and capacities to participate in the economic structure that supports human existence. They may be established in a manner that creates unfairness or fairness in the distribution of opportunities in exercising the self-interests on the principle of individual freedom to choose and participate in all three structures. The fundamental principle of the individual freedom to chose must be seen as a socially constrained freedom, to the extent to which it is in the *allowable space of freedom* and the *ability space of choice*. The allowable *space of freedom* is defined by the rules and regulations as abstracted in the power relations of law and politics. The *ability space* of choice is in turn defined by politico-material relations as may be abstracted from the power relations of economics and politics of human actions in the social decision-choice space. The laws, rules and regulations may be crafted and implemented in such a way by the decision-making core to abstract rent to enhance revenue for governance in the name of the government. The government as an institution resides in good-evil duality. A government cannot be claimed to be absolutely good or absolutely evil. It is a good-evil structure in a continuum where the position in the continuum varies depending on the collective personality of the social decision-making core and the character of the governing.

All individuals in the political economy can refrain from exercising their rights and opportunities to participate in the political and legal structures and still have the opportunities to maintain their human existences to the extent to which they can exercise these opportunities in the economic structure. Nobody can survive for any reasonable length of time when she or he is deprived of all the opportunities to participate in the economic structure. The economic structure is the foundation of life that engulfs all decision-choice activities. The political structure is the foundation of decision-making power that may be acquired by an individual or a group. The legal structure is the foundation of social stability in the three structures in the sense of mediating and helping to resolve conflicts in the three structures in the political economy under the management and administration of the social decision-making core. The political and legal structures define the institutional parameters and the allowable action space in the economic structure that supports social existence. They also set the boundaries of *resource-income-wealth distribution* in the economic structure. In other words, the political and legal structures are constraints on the activities in the economic structure. In this framework of social organization, it is the existence of the potential redistribution effects of wealth and income that motivate rent-seeking behavior by decision agents operating on self-interest.

The existence of the political and legal structures, through the national-interest setting and goal-objective formation may be used by the social decision-making core to generate rent-seeking activities which are then harvested in the economic structure during the implementations. The sequences of the rent-seeking process and the possible corrupt practices are such that national interest and the goal-objective set are created in the political structure and protected in the legal structure which then protects the rent-seeking environment and the rent-seekers under the laws which regulate individual and collective behaviors in the economic structure. The outcome in this process depends on the nature of power distribution that is created in the politico-legal structures in the sense of who controls the institutions of government. Here, the more informed, knowledgeable and resource-powerful individuals, working on the principle of self-interest and on the fundamental postulate of individual freedom, shape the rent-seeking environment through their influence in the setting of the national interest, rule making and formation of national goals and objectives. It is here that business firms and bureaucratic capitalists acquire an undue influence in the political economy, where the members in the social decision-making core are selected by means of some democratic collective decision-choice system.

The irony of this political-law-economic integrated structure is that an organizational attempt on the principle of electoral democracy to protect the weak from the strong in the uncontrolled competitive jungle ends up in protecting the strong against the weak. A question logically arises: Is an individual better off in the uncontrolled competitive environment of the market system or better off in an organized competitive environment of the market system where there may be possible checks and balances in the interest of controls on individual and collective behavior? Here emerges the individual-community duality in a continuum and its behavioral dynamics in the space of freedom. How much freedom should the individual have on the basis of individual self-interest and how much freedom should the collective be accorded on the basis of collective self-interest in defining the social goal-objective set and the implementation of the elements? How should the individual and collective self-interests be related and structured in either a controlled or uncontrolled political economy for social stability and progress? An interesting question follows if we select the latter as primacy (controls) to the former (uncontrolled). Should the markets, which emerge in the economic, political and legal structures, in the organized competitive environment be regulated, and to what extent should it be regulated and by whom? The answers to these questions depend on the degree to which the protection is provided for the strong by the integrated political-legal-economic structure. The political and legal structures may provide the framework for all on the basis of individual self-interest and the fundamental moral postulate of individual freedom to exercise this freedom in the economic structure and the markets it contains.

In this framework, there are individual conflicts in freedom as well as conflicts in the individual-community duality. The practice of individual interest, on the fundamental ethical principle of individual freedom, is such that a lack of regulation will lead to the destruction of the community interest since the practice will be based on the crude notion of *each for himself and God for us all* rather than on the enlightened self-interest of *each for all and all for each*. This is what is known in economic theory as perfectly competitive markets without government in the social decision-choice process. A problem arises as to how does the system reconcile conflicts among individual freedoms and interests as well as individual-community conflicts in freedom and interests in the space of preferences without violence in the three markets and with stability in the social set-up? The answer has always been found in the establishment of central authority called the government with governmental regulations in all three aggregate markets of politics, law and economics. A paradox arises. As regulations are introduced, the exercise of self-interest on the basis of fundamental moral postulate of individual freedom forces an individual to search for rent-seeking areas and to take advantage of legal and illegal loopholes in the regulatory process in all the market.

To close the loopholes calls for more government regulations. More government regulations widen the environment for rent-seeking. We shall refer to the enforcement of the regulatory process as *intensive regulation* and the increasing number of regulations and laws as *extensive regulation*. Regulation-widening increases extensive regulation and creates opportunities for rent-seeking and increasing rent-seeking. Decreasing regulation-deepening decreases intensive regulation and creates conditions for increasing rent-harvesting for the individuals who are pursuing their self-interest on the basis of institutions that are promoting individual freedom without sufficient protection for the collective interest and collective freedom.

Regulation widening, given regulation intensity, or regulation deepening, given, extensive regulation, increases the role of government in the social organism and increases the controls of the limits of individuals in exercising their self-interests within the collective interest and freedom. An increasing regulation widening and deepening continuously reconciles the conflicting priorities of collective interests and freedom with those of the individuals in the social set-up of the political economy, and thus works to reduce rent-seeking activities where such a regulatory system is dynamic and created with social welfare as its criterion in the creation and management of the regulatory regime for the political economy. A decrease in regulation-widening, given regulation intensity, or a decrease in regulation deepening, given regulation widening reduces the role of government in the social organism, and decreases the controls on the limits of individuals in exercising their self-interests within the collective interest and freedom. Decreasing regulation-widening and decreasing regulation-deepening continuously assert the priority of individuals' interests and freedoms over those of the collective in the social set up of the political economy, and thus work to increase rent-seeking activities. Regulation-widening can easily be weakened by decreasing regulation-deepening. This is a case where there are many rules and laws without enforcements. Alternatively, the existing regulations can be strengthened by regulation-deepening which then decreases rent-seeking opportunities by closing enforcement loopholes. The legal structure and its effectiveness are reflected on its extensive and intensive regulatory regime.

The processes of regulation deepening and widening are intimately connected to the national interest definition and goal-objective formation through implementation. The rent-creation, rent-protection and rent-harvesting outcomes of the regulatory process depend on who participate to define national interests and the social goal-objective set. They further depend on the degree of political influence that individuals or groups of individuals can acquire to affect the democratic decision-choice process. The political influence defines what is to be regulated and what regulation rules are to be enforced to affect the market efficiency and possible deviations from the economist perfectly competitive market structure, its imputations and resulting income distribution. Such influence is not social-system specific. However, the manner in which such political influence is acquired and exercised is social-system specific where the applicable boundaries are culturally defined.

Given the role of regulations in the social stability of the political economy, we may raise a question as to whether there is an optimal regulatory regime in the political economy. If there is, then what are the conditions for optimality and how do we create it? Is the optimal regulatory regime the same or does it vary over different political economies? It seems that the conditions of optimality should be related to the size of the potential rent-seeking as seen in terms of the social cost-benefit configuration of the regulations and laws in relation to the cost-benefit of the individual interests and freedoms, relative to the collective cost-benefit of social stability, harmony and general systemic risk of the institutions of decision-making in the political economy. Each social state may have a different optimal regulatory regime that can sustain the social stability, harmony and the corresponding systemic risk. The changes in the social state may also require changes of the optimal regulatory regime. The optimal regulatory regime besides being politico-economic state specific is also specific to different nations with different institutional arrangements. Each nation has a different culture of tolerance and decision and manner in which individual-community conflicts in preferences are resolved to reduce the associated systemic risk. The optimality of any regulatory regime imposes a temporary social equilibrium that is dependent on the decision-choice behavior of the social decision-making core and its social vision, conception of the national interest and required social goal-objective set given the domestic resource endowment. A question may be put to the anti-regulatory advocates. Should the provision of the regulatory services be left to the private sector for profit since the private sector is claimed by them to be more efficient?

The whole analytical form of the relational structure of the controlled and the controlling may be presented in the logic of pyramidal geometry that shows the interactive modes of individual and collective freedoms, the principal and agent and the center of the regulatory regime. It is the existence the set of private-public sector combinations defining mixed economies that the possibility of rent-seeking arises in the various political economies. The size of the rent-seeking space composed of rent-creation, protection and harvesting will vary over the private-public-sector efficiency frontier. It will also vary within the duality of the individual-collective freedom. As such, attention will be directed to examine the associated conditions over the private-public-sector efficiency frontier and how it may be related to Marxian socialism and Schumpeterian socialism as politico-economic duality with a continuum.

Chapter 3
Rent-Seeking Activities in Schumpeterian and Marxian Socio-Political Dynamics

As we have discussed in Chapters one and two of this monograph, the social system is welded together by institutions of the three blocks of economic, political and legal structures. Each of these structures has its own identity for recognition and yet is connected to each other to establish the unity of the political economy as a complex self-improving and self-transformation system. The social system must be seen as an automatic control system where the automatic controllers are linked to the decision-choice system. The economic structure is the foundation of life; the political structure is the foundation of social decision-making power and; the legal structure is the foundation of the social control and stability regarding the individual and collective decision-choice behavior in the three structures. In this respect, the political structure, endowed with the decision-making power, shapes and manages all three structures for coherence, stability and systemic risk associated with the collective decision-choice system of the political economy. At the same time, it is also shaped by the interactive behaviors of the three structures as they affect the social existence. The outcomes and social mandates of the interactive behavior create the environment for rent creation and harvesting. The nature of the rent-seeking environment and the distribution of rent creation and harvesting activities over the population depend on actual and potential political controllers that are used to manage the society over the private-public sector efficiency frontier shown in Figure 3.1.

3.1 Rent-Seeking and the Public-Private Sector Efficiency Frontier

The controllers of the political structure acquire the decision-making power that helps them to shape the three structures, individually and collectively, to define a possible rent-seeking environment usually in the favor of different social class-interests. Such class interests may be related to the members of the decision-making core who create the political and legal controls. The process of acquiring the political-control of the social decision-choice actions generates conflicts and tension within itself. The conflicts and tension provide continual force for socio-economic transformations through qualitative movements and social dynamics of the political economy. These conflicts and tensions are magnified in all social systems organized on the decision-choice criterion of self-interest which is supported by the fundamental ethical principle of individual freedom, where the principle of *law and order* submerges the principle of *freedom and justice*.

It is here that the scholarly works of Marx [R4.76] [R4.77] [R4.78] and Schumpeter [R4.99] [R4.100][R4.101] [R4.103] become instructive in our search for answers to class conflicts and *dilemmas* in democratic social formation of the capitalist type, and its movement toward an economic formation of the socialist type for *class controls* of the decision-making core or the elected body under a defined constitution that mediates the social games of power relations in the political structure. In this respect, we shall turn our brief attention to relate the rent-seeking activities to both Marxian and Schumpeterian political economies in relation to national-interest definition, social goal-objective formation, class-conflict, rent-seeking activities, and income distribution and substitution-transformation processes of the institutional arrangements of the society. All these must be related to the private-public sector duality in closed economic systems. We shall then extend the results to open-economic systems.

Let us examine the impact of the rent-seeking activities on the movement on the private-public efficiency frontier as presented in Figure 3.1. Position A in Figure 3.1 will correspond to a Marxian socialist economy where almost all goods and services are under the provision of the public sector. Here, labor controls the social decision-making core of the government. Position B will correspond to a Schumpeterian socialist economy, where almost all the goods and services are under the provision of the private sector. Here, the bureaucratic capitalist controls the social decision-making core of the government. Under the Marxian socialism $\left(1-\pi_\Pi^r\right)$ is the private-sector proportion of the provision of social goods and services represented by the (EA1). Under the Schumpeterian socialism $\left(1-\beta_\mathbb{P}^r\right)$ is the public-sector proportion of the provision of social goods and services represented by the area (FB1). The point $\mathbf{K}=\left(\pi_\Pi^r,\beta_\mathbb{P}^r\right)$ is not sustainable since $\left(\pi_\Pi^r+\beta_\mathbb{P}^r\right)>1$. For a comparative analysis of the differences and similarities see [R15.16].

It may be noted that, in the general societal organization, the **OC** is an *irreducible private sector proportion* while the **DOFB** is the *irreducible private sector area* proportion of activities in social production. The distance **OD** is an *irreducible public sector proportion* while **COEA** is an *irreducible public sector area* proportion of activities in social production. The rectangle defined by the distance **OD** and **OC** is the non-negotiable area of social production. Every social system has a non-negotiable area of social production that is established by social norms and ideological conditions which allow joint participations of the private and public sectors. We shall begin with rent-seeking in the Schumpeterian political economy and then reflect on the nature of rent-seeking in the Marxian political economy and relate them to the public-private efficiency frontier. The public-private sector efficiency frontier presents to us trade of possibility of institutional arrangements of public and private sectors for the provision of social goods and services. It projects the increasing substitution costs of using the private or public sector to provide social goods and services beyond some defined points give the social prefereces.

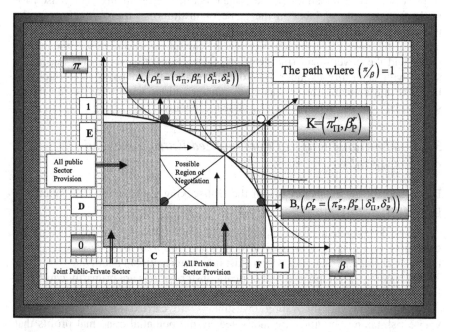

Fig. 3.1 The Relative Positions of Public-Private Sector Duality in Relation to Schumpeterian Socialism and Marxian Socialism with the Private-Public Sector Efficiency for the provision of goods and services in the Political Economy where

π = public-sector proportion, β = private-sector proportion,

$\pi_{\mathbb{P}}^r$ = public-sector reservation proportion by the private sector,

β_{Π}^r = private-sector reservation proportion by the public sector, and

$\pi_{\Pi}^r - \beta_{\mathbb{P}}^r$ = ideological distance and zone of negotiation

3.2 The Schumpeterian Political Economy and the Rent-Seeking Process

To relate the Schumpeterian economy to the rent-seeking process, we shall first define the basic characteristics of the structure of the Schumpeterian political economy and its dynamics. The players in the Schumpeterian dynamics in the political economy are entrepreneurial capitalists, bureaucratic capitalists who represent the business sector and the state that represents the public sector. The resolution of the conflicting preferences of these players generates the qualitative law of motion that transforms the political economy with continual private-sector dominance towards point B in Figure 3.1. We refer to this as the Schumpeterian qualitative law of motion. The labor or the work force does not play any role and hence is completely neglected in the Schumpeterian economy as a relevant force to the institutional dynamics in the private-public sector duality. The reason for neglecting the role of labor in the game of institutional transformations in the private-public sector duality in the political economy is based on the conceptual nature of the power distribution between the capitalists who represent the private sector and the government (the decision-making core) that represents the public sector with neutrality of labor. The dynamics are seen in terms of politics and business that generate the moments of transformations within the private public sector duality in the Schumpeterian political economy.

3.2.1 The Game Stage of Personalized Entrepreneurial Capitalism

Let us specify the characteristics of the rent-seeking game in the Schumpeterian political economy.

a) The development of the capitalist economy is the result of activities of the capitalists as a class which is divided into an *entrepreneurial class* and a *non-entrepreneurial class*, both of which exist in transformational duality.

b) Economic developments and critical transformations are primarily caused by innovation and innovation investments in the key sectors of the economy.

c) Innovation and innovation investments are carried on by entrepreneurs whose risk-taking activities are motivated by the potential abnormal profits that become realized rewards for risk-taking.

d) The innovation-induced abnormal profits invite intense competition from non-entrepreneurial capitalists who enter a particular innovated area with duplicating investments by methods of mimicking to flood the market, deplete abnormal profits and create increased profit-uncertainty and risk for entrepreneurial capital losses. Equilibrium is established with zero abnormal profits and an optimal number of non-entrepreneurial firms.

e) The entrepreneurial class looks for new areas for innovation and innovation investments to create abnormal profits for harvesting. Conditions (c and d) are then repeated.

f) Innovation and innovation investments by the entrepreneurial capitalist class create abnormal profits; competition by the non-entrepreneurial capitalist class eliminates them forcing the creation of a *new combination* by the entrepreneurial capitalist class. The process is what Schumpeter refers to as *creative destruction* (which is referred to here as Schumpeterian *qualitative equation of motion*).

g) The dialectical moment of transformation is thus composed of innovation, abnormal profits and competition between the entrepreneurial class and non-entrepreneurial class while the government, operating through the political and legal structures, maintains fairness in competitions and referees the economic game between the entrepreneurial class and non-entrepreneurial class with given national endowments under the principle of individual interest and the fundamental ethical postulate of individual freedom in the three structures of the political economy. Innovation with innovation investments creates abnormal profits, while competition with non-innovation investments depletes and destroys abnormal profits.

g) The labor or workforce is simply available at the capitalist productive power. Similarly, the members of the labor or work force can transform themselves into either entrepreneurial or non-entrepreneurial capitalists to compete for abnormal or non-abnormal profits.

h) The role of the government is to regulate the market system through the legal structure to promote fair competition and productive social incentives beside the price movements.

The organization of the political economy is provided in Figure 3.2. The markets of the real sector and the financial sector are regulated for effective coordination among the financial and real sector participants for the stability of the political economy. The entrepreneurial capitalist political economy in the Schumpeterian system is characterized by entrepreneurial-non-entrepreneurial duality where innovation and non-innovation investments over profits are the central features of the duality and the driving force of transformation between categories of development. The private sector is the dominant feature of the political economy. Conflicts in the differential capital and profits by the innovators and non-innovators define the dynamics of the Schumpeterian political economy and its transformations.

The central features of entrepreneurial capitalism may be summarized as:

a) Entrepreneurs own and control capital and its business operations.

b) Entrepreneurs run, and manage their businesses (that is, the managerial responsibility rests on the entrepreneurs); there is no separation between ownership and management.

c) Entrepreneurs manage the economy by controlling the physical capital, financial capital, labor employment and income distribution through the markets of real and financial sectors.

d) The real sector is used for the production of goods and services, while the financial sector is used to support the financing innovation, innovation investment and risk-taking in the real sector where the capitalists produce goods and sell them for profit.

e) Entrepreneurs receive capital income as the benefit of ownership. Both profits and losses are the responsibilities of the entrepreneurs and hence are completely *privatized*.

f) Entrepreneurs control the capital and production power but do not control the political power and hence do not control the power to make laws and regulate themselves for their profit-making activities. They also do not necessarily control the financial power. They function under the principle of individual interest and the postulate of individual freedom in the market.

g) Innovation, innovation investment and risk-taking are their tools for profit making. The interactive system is illustrated in Figure 3.2 where the focus of the capitalist is individual profit optimization with the instrument of economic power, and the focus of the government is maximization of fairness of competition and social welfare with the instrument of political power. The government controls the political and the legal structures (powers) while both the entrepreneurial and non-entrepreneurial capitalist classes control the economic structure (power).

h) Rent-seeking, protection and harvesting are not part of the entrepreneurial capitalist political economy. The essential activities of the capitalist class are profit-seeking, profit-creation and profit harvesting under a free-market system in the Schumpeterian *decentralized capitalism* which is also personalized capitalism.

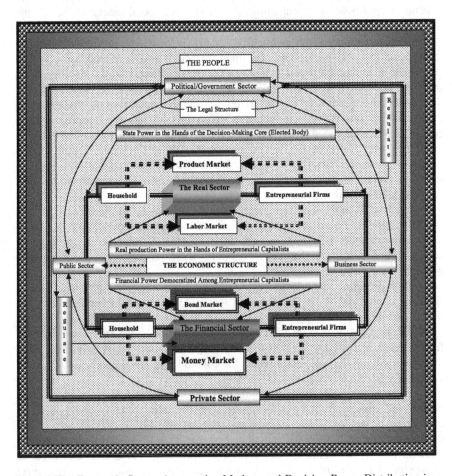

Fig. 3.2 The Economic Space, Aggregative Markets and Decision-Power Distribution in a Domestic Entrepreneurial Capitalist Political Economy

The general structure of entrepreneurial capitalism, the roles of the state, entrepreneurial and non-entrepreneurial capitalist classes and the power distribution are shown in terms of a cognitive geometry of knowing in Figure 3.3. The distribution of social power is such that the economic power is substantially controlled by the private sector composed of entrepreneurial and non-entrepreneurial capitalists. The state, through the decision-making core, controls the political and the legal powers to regulate the market system and the nature of participation in the political economy to ensure fairness in competition among economic participants and how this fairness will ensure market democracy. The government or the public sector plays an insignificant role in the provision of goods and services in fulfilling the social vision, and the implementation of the elements in the social goal-objective set to support the achievement of the national interest and social vision. In this frame, the set of private institutions dominates the set of public institutions at least in the provision of goods and services.

From the viewpoint of qualitative dynamics, every state of the political economy is seen as belonging to a politico-economic category. A movement from one state to the other is thus categorial conversion of social states that allows the political economy to lose the qualitative characteristics of the previous state and acquire new qualitative characteristics that establish a distinction. The conditions

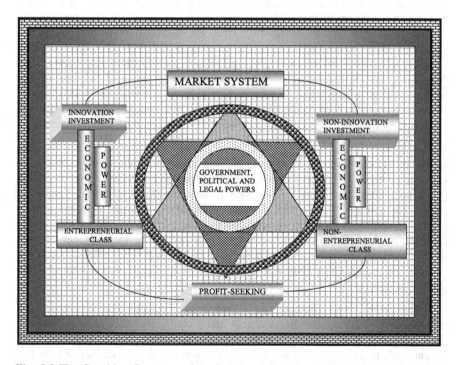

Fig. 3.3 The Cognitive Geometry of the Relational Structures of the State as the Public Sector, the Market System and the Power Distribution in the Entrepreneurial Political Economy

that are required to make the inter-state movement possible are called *convertibility conditions* which are generated within the social state in the sense that they are created by intrastate forces of motion. The existence of the social convertibility conditions is necessary but not sufficient for an inter-state transformation. The sufficiency requires the creation of a force of inter-state motion that is called a *categorial moment* that will bring about the needed categorial conversion from state to state. In Schumpeterian transformation dynamics, the convertibility conditions are: a) existence of markets, b) conditions of self-interest, c) incentive conditions of profit-making, d) intense competition for profit e) entrepreneurial capitalists, f) non-entrepreneurial capitalists, and g) a mediating role of the state using the political and legal structures. The a), b), and c) constitute the convertibility conditions while d) e), f) and g) generate the needed categorial moment for moving the Schumpeterian entrepreneurial capitalist political economy through states. The private-sector interest is controlled by the innovative and non-innovative entrepreneurial capitalists.

3.2.2 The Rise of Bureaucratic Capitalism in the Schumpeterian Political Economy

Let us turn our attention to the evolving nature of the capitalist class, the market system, the structure of the profit-seeking activities, the role of the government and the changing power distribution over the political, legal and economic structures as the conflicts between the entrepreneurial and non-entrepreneurial capitalist classes intensify to require increasing governmental activities in the Schumpeterian entrepreneurial political economy. The essential elements of the evolving nature of the initial Schumpeterian political economy are stated below. The process is that both the entrepreneurial and non-entrepreneurial classes are transformed by the profit-seeking process into a bureaucratic capitalist class to create *Schumpeterian bureaucratic capitalism.*

a) The process to eliminate conditions of competition between the entrepreneurial class and non-entrepreneurial class over abnormal profits that create increased profit uncertainty and capital risk, leads to the rise of a bureaucratic capitalist class. The bureaucratic capitalist class establishes *centralized bureaucratic capitalism* that is composed of monopoly capital, mega-groups of businesses operated by skillful technocrats and bureaucrats through the methods and toolbox of takeovers and buyouts in horizontal and vertical integration in order to reduce competition, profit uncertainty and capital risk which constitute the core of their path to increase profits.

b) The centralized bureaucratic capitalism replaces the *personalized entrepreneurial capitalism.* Bureaucratic and technocratic management replaces entrepreneurial management. The entrepreneurial competition is replaced with bureaucratic private-market restrictions and predatory competition that destroy the creative component of the entrepreneurial competition in the creation of the new and the replacement of the old. It has little or no respect to public-sector interest and social welfare. It has more respect for the interest in increasing the private business sector's profit and bureaucratic income. The principles of individual interest and the fundamental ethical postulate of individual freedom are replaced with the principle of bureaucratic interest and the fundamental ethical postulate of bureaucratic freedom for the bureaucrats and technocrats as a class in the private sector of the political economy. The non-bureaucrats composed of individual entrepreneurs, non-entrepreneurs and workers are reduced to subservient and supporting roles with their creative forces reduced to nothingness in the intra and inter-state transformations where the government still maintains its role of creating market fairness in competition which is no longer consistent with the existing institutions of the entrepreneurial market system. The market becomes restricted by decision-choice activities of the bureaucratic capitalists, reducing competition creating unfairness and destroying the market democracy. A conflict arises between the government whose goal is to maintain fairness in the competition as well as workable market democracy, and the bureaucratic capitalist class whose goal is to restrict competition and market democracy to enhance its profits.

c) New institutional arrangements with a new value structure are established to define a new *power relation,* a new framework for the socio-economic conflicts between the business sector represented by the centralized mega-bureaucracy and the public sector represented by the state and managed by the elected body (the social decision-making core). In other words, the entrepreneurial-non-entrepreneurial duality is transformed into an intense private-public sector duality with different dynamics.

d) The game of institutional dynamics between the entrepreneurial capitalist class and the non-entrepreneurial capitalists is also transformed into a new game between the bureaucratic capitalist class and the government where the bureaucratic capitalist seeks to control the government through the political structure for profit-making.

e) The game introduces a new condition of categorial convertibility with a changing categorial moment where the role of the entrepreneurial capitalist is reduced to practically a non-force or nothingness.

3.2.3 The State, Bureaucratic Capitalism and the Rent-Seeking Process

The essential structure of the relationships between the government and the Schumpeterian bureaucratic capitalism and the rent-seeking process may be outlined in in the following steps.

a) To defend and maintain its existence, the centralized bureaucratic capitalism, operates by continual size enlargements through takeovers of competitive firms by building vertical and horizontal integrations, acquires market power, restricts market competition, absolves creative talents and innovators, and enhances abnormal profits that allow it to acquire *social power*. The social power is composed of economic, political and legal powers, in order to override the *power of the state* that operates in the political structure to maintain fairness and referee the competitive market game through the legal structure.

b) The nature of bureaucratic capitalism is to use its market power to destroy or restrict market competition, competitive innovations and affect intra and inter structures of income distributions in all sectors and in the political economy. The nature of the State is to exercise its power through the decision-making core to create competition and to ensure fairness in the legal and economic spaces in order to optimize social welfare and not private profits. This requires taking into account the welfare implications of the income distribution of income recipients. A classic duality emerges to give rise to an intense conflict between the large business organizations that represent the interests of the business sector for profits and increasing profits without the concern of the overall welfare of the nation, and the State which represents the interests of the public sector where the role of labor in the power relation is reduced to nothingness, and where the conditions of democratic decision-making are completely suppressed by nominal income power in the economic structure. The conflict is in relation to defining the relative importance of interests between the private and public sectors. Here, the bureaucratic capitalists seek to define public and national interests in terms of private-business interests and the State seeks to define private-business interests in terms of public and national interests. The conflict in the private-public sector interests may be seen in terms of the proportional distributions that affect education, housing, health, law, justice, security and others.

c) The conflict in the public-private-sector duality manifests itself in the general social decision-choice space defined in terms of differential individual preferences. The members of the social decision-making core representing the state (the government) as the agent, reasoning from the principle of *creative component* of competition and the postulate of individual freedom as it relates to collective freedom under an accepted ideology, seek to regulate and control the competition and fairness in the economic structure through the legal structure in accord with the perceived public and national interests. The members of centralized bureaucratic capitalism, reasoning from the perception of the *destructive component* of profits in competition, design strategies to evade the rules and regulations in order to manipulate competitive conditions in the market as well as conditions of individual interests and freedom, to generate abnormal profits through an indirect destruction of market democracy. The acquisition of abnormal profits allows the bureaucratic capitalists to maintain their social power without regard to collective freedom and general social welfare. Collective freedom and general welfare of the society are not their goals, and bureaucratic capitalists will pursue every strategy to enhance profits and bureaucratic income even when the strategy is antagonistic to the national interest and social welfare of the State.

d) It is the conflict between the *State political power* to legislate rules and regulations to change the legal structure to affect the economic structure, and the *bureaucratic economic power* of the bureaucratic capitalists to evade the rules and regulations that introduces into the bureaucratic capitalism a new phenomenon of active *rent-seeking* in the political economy. From the rise of the phenomenon of rent-seeking, emerges two new classes of the *innovating rent-seeking bureaucratic class* and *non-innovating rent-seeking bureaucratic class* to replace the entrepreneurial capitalists and non-entrepreneurial capitalists respectively. A new socio-political game emerges with its institutional transformation dynamics. It is a power game between the State and the bureaucratic capitalist class in the collective decision space of the political economy.

e) The power relation between the innovation and non-innovating classes in the entrepreneurial political economy is transformed into a power relation between the bureaucratic capitalist class and the State in the bureaucratic political economy.

The innovating rent-seeking class and non-innovating rent-seeking class act as a unit to maintain the established bureaucratic capitalism. They also act to maintain the interests of their bureaucratic capitalist class that represents the interests of the private sector. The innovating rent-seeking bureaucratic class assumes the role equivalent to the entrepreneurial class in the sense of creating *rent-seeking innovations* and *rent-seeking innovation investments*. This is comparable to the entrepreneurial class that creates innovation and innovation investments. The non-innovating rent-seeking bureaucratic class assumes the role equivalent to the non-entrepreneurial class in the sense of entering the rent-seeking space to deplete the rent to zero. Unlike the entrepreneurial capitalist political economy, the conflict is not between the innovating rent-seeking bureaucratic class and non-innovating rent-seeking bureaucratic class but rather between the bureaucratic class and the State. The game is between two powers, the *economic power* held by the bureaucratic class and the *political power* held by the state to manipulate the legal power in the legal structure. The conflict is thus transformed from intra-structural one of the economic structure into inter-structural ones involving economic, political and legal structures. The guiding principle of the use of economic power by the bureaucratic class in the democratic decision-making system is the principle of *bureaucratic self-interest* and the fundamental ethical postulate of *bureaucratic freedom* under the guise of private-sector interest and market capitalism (where in the case of the United States of America, corporations acquire a status as persons (citizens) with political rights as may be specified). The guiding principle of the use of political power by the State through the decision-making core in the democratic decision-choice system is the principle of individual self-interest and fundamental ethical postulate of individual freedom under the guise of public sector interest and ideology of market capitalism under fair competition with the instruments of law-making.

The initial attempt by the members of bureaucratic capitalism to play the power game is to enter the market in the legal structure that has been set in place by the state which also defines the rules and regulations of the legal market and sub-markets. Here, the legal structure with its legal power is viewed as neutral between the economic power and political power. The mega-corporations individually enter into the legal structure to fight against the rules and regulations that they believe function to reduce their competitive edge and restrict their profits. In the legal market, however, an individual corporation, no matter how big, is ineffective in the power game and is always losing both the power and the economic games to the State. The reason is that the individual-corporation's specific game efforts and hence the principle of individual corporate self-interests and freedoms work against the individual bureaucratic capitalist since it has no support from within the bureaucratic capitalist class. It functions under the notion that "each for itself and God for us all" in the power game of political power against the economic power. The game is expensive in that the total individual bureaucratic costs seem to outweigh the total individual bureaucrat benefits in the game. The cost-benefit implication is such that attempts to reduce costs and increase profits end up to increase costs and reduce profits and business goodwill, as part or all the costs are passed on to the consuming public. Whenever a win is registered by a bureaucratic capitalist, the loop-holes in the legal structure are tightened by both extensive and intensive regulations.

It becomes clear that the game against the State in the legal structure produces an unsettling problem, and hence it is not the right place to wage the power game. The reason may be found in the notion that the members of the bureaucratic class do not control the apparatus that creates the rules and regulations which affect the behavior of the economic decision-choice agents as well as the economic structure which they effectively control. The legal structure is to function as neutral in the sense of mediation between the economic and political structures. This is the *principle of organizational neutrality* of the legal structure and the administration of the laws, rules and regulations is there to manage for justice and fairness but not to create. However, the administration of the laws, rules and regulations has integrity to the extent to which the legal administrative personnel are honest with integrity. In this respect and from the viewpoint of the bureaucratic class, the strategies of the game must be transcribed from the legal structure to the political structure that provides the power to change the behavior of the legal structure and its administration and the corresponding personnel. In the final analysis, the losers of the power game between the bureaucratic capitalists and the State are the consuming and income earning public in terms of higher prices and taxes. Let us keep in mind that distribution of income is closed among income recipients at any production-consumption time. That is to say that the distribution of costs and benefits is closed among economic agents in the political economy in the sense that at any production-consumption time, there is no undistributed income.

The rent-seeking class is aware that in order to continue to generate abnormal profits in the economic structure it must directly or indirectly control the political structure that harbors the social decision-making power to create laws, rules and regulations and enforce them with the state power, in a sense, to have direct or indirect control of the power of the state and use it to act in their favor. It knows that after the social decision-making core has been established by the electoral process, the members are free to pursue any social goal-objective element and national interest under the constraints of the legal structure and the members' preferences. The individual preferences in the political economy can be shaped by manipulating the information structure to create a deceptive information structure to create information asymmetry, in order to reshape the decision-choice actions of the principal by the agent which is the government. This deceptive information structure may be directed to favor the public-sector interest or the private-sector interest to change the cost-benefit distribution in the political economy through the activities in the political but not in the legal structure. The incentive to manufacture a deceptive social information structure resides in the actual or potential social cost-benefit distribution among the collective decision-choice agents in the political economy.

The strategies of the power game between the bureaucratic capitalist class and the state controlled by the social decision-making core, is shifted from the legal structure to the political space where the power game is played in the political structure in order to control the political economy and social decision-choice process economy through the rule-making activities in the legal structure. The bureaucratic capitalist class controls the economic power in the economic structure through its ownership of the productive capital. The social decision-making core, by inheriting the institution of the government, controls the legal power through its control of the political power. The bureaucratic capitalist class enters the game armed with economic power. It knows that the foundation of the political economy is the economic power and that when this economic power is strategically and tactically used; it can bring instabilities into the political structure through its behavior in the economic structure and destroy the ruling decision-making core for a change. The social decision-making core, the representative government on the other hand, enters the game armed with State power, where its economic power rests on the generosity of the bureaucratic economic behavior that also controls the revenues through employment of factors. The reward of the game is to acquire the legal power to regulate the activities of the social decision-choice space through the making and administration of laws and regulations in the political economy. Here, an important conflict arises, in the sense that the bureaucratic capitalist class holds the responsibility and the power to bring into fruition the elements in the goal objective set that will support the social vision and national interest which may not meet the preferences of this capitalist, while at the same time it is being controlled by the state leading to a classic conflict game.

To play the game, the bureaucratic class helps to create political market and sub-markets and shapes them with their economic power. It sees this market as composed of sellers and buyers as they are analytically used in the economic market in which it controls and operates in the real sector. The output in this market for buying and selling is *influence* to shape the formation of the social goal-objective set, implementation of the elements of the social goal-objective set, the legal structure and the national domestic and international interests. The sellers of the influence are the elected members or the members of the decision-making core or their representatives. The buyers are the general public and the bureaucratic capitalists. The price is determined by supply and demand where the income elasticity of influence depends on the size of possible rent in relation to regulation widening and deepening. Most individuals in the general population are priced out of this political market leaving the bureaucratic capitalist class as a monopsony in the political market where it faces many sellers of influence from the decision-making core, especially if the bureaucratic capitalists and their enterprises are viewed as citizens comparable to the members with full constitutional rights.

The bureaucratic production enterprise is accorded all rights where it can influence the allocation of resources to the democratic decision-choice system but cannot directly participate in the decision-choice actions in the social decision-making core. The manner in which these resources are distributed may come to corrupt the democratic collective decision-choice system. Furthermore, if the legal process imposes limitational structures on contributions, then a problem will arise in terms of double counting where the members of the bureaucratic class contribute in their own names as well as the names of their enterprise. These contributions are seen as costs of production and passed on to the general public in terms of increased prices or reduction in real wages and non-bureaucratic income. The bureaucratic capitalists with common interests and economic power identify rent-seeking innovations in the implementation of the elements in the goal-objective set. They then buy influence through the use of political contributions and other forms as allowed by the existing legal framework to shape social preferences and the direction of the democratic decision-choice process.

The costs of this transaction are mostly undertaken by bureaucratic rent-seekers on behalf of the bureaucratic enterprise as a first step in rent-seeking innovation investment. They develop political marketers and political action lobbyists within the confines as imposed by the legal structure to take advantage of the influence. The job of the marketers and lobbyists, who represent the bureaucratic class, is to sell the decision-choice preferences of the bureaucratic class regarding the design of regulations, rules and laws that create rent-seeking environments for bureaucrat capitalism. The initial rent-seeking innovation investments introduce the preferences of the bureaucratic class as constraints on the decision-choice behaviors of some influential members of the decision-making core whose influence has been secured and hence socially compromised. Sometimes the influence may be so strong that the lobbyists and the members of the bureaucratic class develop and write the laws, rules and regulations that affect the economic decision-choice space involving the market structure, production, consumption and income distribution.

In other words, the bureaucratic class indirectly assumes the functions of the legislature and shapes the legal structure in the favor of the viability of bureaucratic capitalism. In this way, the bureaucratic capitalist class is brought into unity of purpose with the social decision-making core that manages the governance of the society. The government becomes indistinguishable from bureaucratic capitalism and the bureaucratic capitalist class becomes indistinguishable from the social decision-making core. The State becomes indistinguishable from bureaucratic capitalism in the social practice where the private interests of the bureaucratic class are defined as similar to the public interests, and the interests of the decision-making core are stated in terms of bureaucratic capitalism and ideologically presented to the society as the public interests with a slogan *what is good for the private sector is good for the public sector*. Given the rules, regulations and laws that preserve the rent-seeking and rent-creation process, a second rent-seeking innovation investment is undertaken to harvest the rent that goes to maintain bureaucratic abnormal profits in the economic structure. In this way, the entrepreneurial abnormal profits are transformed into bureaucratic abnormal profits.

3.3 Bureaucratic Capitalism, Socialism and the Schumpeterian Political Economy

The stage where the bureaucratic capitalist class is brought into unity with the decision-making core and where it controls the political structure is called the *Schumpeterian socialism* with its unique characteristics substantially different from *Marxian socialism*. The Schumpeterian socialism is *capitalists' socialism* controlled by the State for the bureaucratic capitalist class at point B in Figure 3.1 where the means of production is in the hands of the private sector. The bureaucratic capitalist class controls the economic, political and legal powers. Marxian socialism is *workers' socialism* controlled by the State for the workers and the public at point A (also in Figure 3.1) where the means of production is in the hands of the State. The state thus controls the political, legal and economic powers. On the road to Schumpeterian socialism, capitalists own the means of production and work to control State power to serve their class interests and maintain private ownership of means of production, privatizing profits and socializing losses. It is an economic system where the means of production is directly controlled by the bureaucratic capitalist class and State power is directly or indirectly controlled by the bureaucratic capitalist class. In this case, the bureaucratic capitalist class comes to control the economic, political and legal powers of the socialist economy in Schumpeterian institutional equilibrium, at least for now.

At the level of organization, the Schumpeterian socialist state is such that there are institutionally divided private and public sectors with a government that holds the political power to govern. At the level of the political structure, the political power is vested in the decision-making core by some form of democratic decision-choice process called election on the principle of majority decisions in the political market. State power, however, is held in the hands of the bureaucratic capitalist class through its direct or indirect control of the elected body that constitutes the social decision-making core in order to advance its interests. In fact, some of these members may have come from the bureaucratic capitalist class. The control mechanism comes through the buying of influence in the political market place in three steps. The first step is through extensive costly advertising to affect the voting outcomes in constituting the decision-making core where the members, favorable to the interest of the bureaucratic capitalist class, are elected. The second step is through direct or indirect purchase of influence from the members of the elected body. The third step is through the use of the acquired influence to shape the rules and regulations that must enter as part of the legal structure. The bureaucratic capitalism is then completely transformed into *political oligarchy* controlled from the economic structure by a small but powerful bureaucratic capitalist class. In this transformation, the behavior of the economic structure is no

longer controlled by rules and regulation from the political structure as in the case
of an entrepreneurial political economy of the Schumpeterian type.

The legitimacy of the political oligarchy is derived from a complex electoral
process with a deceptive information structure of mass disinformation,
misinformation and support of the constitution. In this process, and with mass
acceptance of the elections as a resolution of conflicts in the democratic collective
decision-choice space without violence in selecting the decision-making core, the
political oligarchy acquires the status of *democratic oligarchy* where the
bureaucratic capitalist class forms a natural aristocracy and the social system is
projected as a *representative democracy* or government by the people, of the
people and for the people as an ideological principle of mass deception and
confusion in the democratic decision-choice space. The resulting transformation is
that the democratic oligarchy becomes domestic colonialism where the colonized
and governed are the workers, broadly defined, and the colonizers are the
bureaucratic capitalist class that constitutes the social aristocracy where the
government is by the people through electoral process and for the democratic
oligarchy by the constitution. Democracy acquires the character of a mirage in the
cognition of the people and this democratic mirage is maintained by walls of
ideology that constitute its protective belts. There are two important qualitative
characteristics going on in the bureaucratic capitalist political economy. It restricts
democracy in the economic structure, and with the restrictions of market
democracy, it works to restrict democracy in the political and legal structures. The
democratic mirage, as a collective cognitive illusion and the social stability of the
bureaucratic capitalism are directly or indirectly maintained by the principle
of law and order which is then supported by institutions and instruments of
enforcement. The effective utilization of the institutions and instruments of
enforcement is then supported by information collection through institutions
National Security State and the use of instruments of severance to destroy the
property of private information. The bureaucratic capitalism, National Security
State, the principle of law and order and the instrument of enforcement violate all
the characteristics of the principle of freedom and justice demanded by true
democracy.

In this framework, the social adjustment coefficient $\alpha_\mathbb{P}$, for the transformation
dynamic equation, depends on the power relation between the bureaucratic
capitalist class and the social decision-making core that occupies the State. We
may specify the equation of motion for the categorial conversion as:

$$\rho_\mathbb{P} = \beta_\mathbb{P}^r + \alpha_\mathbb{P}\left(\beta_\Pi^r - \pi_\mathbb{P}^r\right), \ \beta_\mathbb{P}^r > \beta_\Pi^r \text{ and } \alpha_\mathbb{P} = \alpha_\mathbb{P}\left(\sigma_S, \sigma_B\right) \in (0,1) \quad (3.1)$$

The values β and π are as defined in Figure 3.1, where the value, σ_S is a
measure of the power of the state while the value σ_B is a measure of the powers
of the bureaucratic capitalist class in the political economy. The speed and the cost
of adjustment over the private-public sector efficiency curve is then defined by

the behavior of $\alpha_{\mathbb{P}}(\sigma_S, \sigma_B) \in (0,1)$ while the speed of adjustment depends on the political adjustment costs that must be mapped into the economic space. The political adjustment costs in turn depend on the conditions of the legal structure. The study of the behavior of these political adjustment costs will reveal the nature of the political dynamics as power adjustments are made to control the legal power in the social set up. There is an inverse relationship between State power and the power of the bureaucratic capitalist class in terms of marginal rates of power substitution in the Schumpeterian political economy. The marginal cost of the acquisition of more political power increases as more and more power is acquired by the bureaucratic capitalist class.

At the level of the legal structure, the legislative power is constitutionally vested in the decision-making core that becomes directly or indirectly controlled by the bureaucratic capitalist class after a successful power game. The control of the political structure through the decision-making core is used to design laws, regulations and rules in favor of private sector interests in which the bureaucratic capitalist class dominates by holding the economic power. Besides the direct or indirect control over the social decision-making core, the dominance in the economic structure is expressed in two forms in exercising the economic power over labor and over small producers. By controlling the decision-making core it effectively controls the legislative process, the rule of law and the legal enforcement process. In other words, with the economic power, the bureaucratic capitalist class comes to control the political power which allows it to redesign the legal structure for control of the legal power, to further increase its economic power over the social setup. By well designed but dubious strategies, the members of this class redirect tax revenues to their advantage by shifting the social costs of their private operations to the public sector.

At the level of the economic structure, the means of production (capital) is privately owned and controlled in such a way that the economic system is dominated by the bureaucratic capitalist class who controls the economic power of the nation, and hence controls the livelihood of labor through extensive and intensive settled exploitation. The only avenue of social relief in the political economy is through the legal structure which is now controlled directly or indirectly by the bureaucratic capitalist class. The economic power is made up of *real production power* in the real sector where goods and services are produced, and *financial power* that resides in the financial sector where credits, IOUs and other financial instruments are produced either in support of real investment and consumption or in support of financial games. Profits are privatized through the manipulations of the political and legal structures by the bureaucratic capitalist class and losses are socialized with increasing privatization of public services and the duties of the government that are substantially financed by wage taxation, to further enhance bureaucratic profits, executive incomes and reductions of bureaucratic losses. With an effective direct and indirect control of the government through the control of the social decision-making core, the members of the bureaucratic capitalist class work to privatize everything in their way for rent-seeking and profit making. To them and their academic supporting theories of the political economy, every social problem is solvable by simple market imputations and profit-making incentives. They pay themselves high salaries as well as give themselves big bonuses. State power is captured directly and indirectly to use tax revenue to finance losses associated with inefficient bureaucratic capitalist management and risk-taking behavior (for extensive discussions on uncertainty, expectations and risk see [R7.4] [R7.15] [R15.1] [R11.26] [R15.21). It seeks to reduce its tax commitments and increase the burden of taxes of the non-bureaucratic capitalist class through dubious manipulations of the legal and political structures. The process is to change the incentive structure and the income-wealth distribution in the political economy. The result is extreme income-wealth distortions in the income-wealth distribution where the highest wealth-income proportion is in the hands of a small proportion of the population creating an expanding area of the poverty space. The legal structure is used to close possible avenues of public relief or *voucherizing* the possible public services of such relief to the private sector for more profit making.

At the level of domestic education and culture, the bureaucratic capitalist class controls the instruments of education, colonizes all the academic institutions of higher education and research and forces them to teach the refined ideology of bureaucratic socialism or the democratic oligarchy under the pretext of democracy. Scientific discovery and technological innovations are directed to strengthen the power of the bureaucratic capitalist class where, in fact, the military and law enforcement are organized to depend on the interest of the bureaucratic capitalist class. It can do and does this because the members now effectively control the two powers of economics and politics that regulate social stability of the nation through the legal structure. At the level of domestic social stability and tolerance of democratic oligarchy and its bureaucratic exploitation, ideological caveats are created from the initial conditions of entrepreneurial capitalism, where the guiding principles are individual interest and ethical values of individual freedom.

The bureaucratic interests, profit-making incentives and freedom are created as a social ideology and then encapsulated as the principles of social efficiency. At the level of the political structure, these principles are promoted as the principles of individual interests and individual freedom under democratic-choice system. At the level of the legal structure, they are promoted as principles of law, order and personal accountability. At the level of the economic structure, they are promoted as the principles of individual freedom to choose, and personal responsibility in participating in the production-exchange-consumption game in the market system that holds individual interests. Collective interests and freedom are called into action as individual duty and responsibility in terms of the principle of patriotism when the foundational interests of the oligarchy are threatened. The government, controlled by the bureaucratic class, is marketed to the masses as a democratic government that serves them and the public interests. The government is promoted as government by the people through the democratic collective decision-choice system, government of the people in the sense of ownership, and then government for the people in the sense of the service for the people. The State power is, however, used to protect the bureaucratic capitalist class, business aristocracy and their interests while it is simultaneously used to restrain labors' opposition through complex maize of laws that have been accepted by the masses as immutable self-evident truths for their protection. The military is organized and used to advance and protect the financial and economic empire of the bureaucratic capitalist class under the pretext of national interests, state security and public safety, where the nation's interests are the same as the interests of the bureaucratic capitalist class, state security is the security of bureaucratic capitalism and public safety is simply the safety of the bureaucratic capitalist class. Transparency of governance becomes a casualty where relevant information to the voting public is classified under the pretext of national security that functions through the instruments of the *deceptive information structure* which is composed of *disinformation* and *misinformation structures*.

At the level of social communications and information sharing, the instruments of mass communication and dissemination of information are acquired and controlled by the bureaucratic capitalist class in the economic space, as part of business operations to promote ideological caveats. Education on mass scale is reduced to mimicry and pure parody for the system's obedience without the development of critical thinking to question the social foundations. Science is promoted to the extent to which it serves the production interest of the bureaucratic capitalist class. In a sense, a new class of ideological slave workers is created to service Schumpeterian socialism where labor is reduced to the same level as capital under the control of bureaucratic capitalists. Labor is reduced to a surrogate capital under the effective control by the business aristocrats. Education materials for learning and publishing materials from research are controlled through a well-crafted financial system of funding that guides the behavior of the refereeing process of the academic elite through the business of publishing under the ideology of *quality control* with rewards and punishment. University professors are co-opted to be adjunct members, or hope to be one of the bureaucratic class through revolving-door policy of research funding to maintain the cognitive padlock to the ideological box where creative thinking is imprisoned within walls of familiar and parody of acceptance. Their home institutions receive endowments, research funding from both the business sector and the government that are effectively controlled by the bureaucratic capitalist class which directly or indirectly defines and controls the areas of research and the enterprise of the knowledge production. State power is used to classify and restrict information to the public thus forcing members of the public to operate in ignorance and in a zone of information sub-optimality without clear understanding of the social system's behavior.

3.4 Creative Destruction, Rent-Seeking and Democratic Oligarchy in Schumpeterian Political Economy

Let us examine in a little comparative detail, rent-seeking as a *creative destruction* of the democratic process through its effects in the economic structure. The economic structure of bureaucratic capitalism, like that of entrepreneurial capitalism, on the aggregate, is composed of the real and financial sectors. The real sector which is the foundation of life is made up of two markets of labor and commodity markets as market aggregates. The financial sector as the store of value and facilitator of exchange among buyers and sellers in the real sector is taken to be composed of the money market, and bond market for debt and other loan-borrowing financial instruments. A working mechanism of the organization of capitalism is useful to the understanding of rent-seeking as a process of creative destruction of democratic social formation under market capitalism, individual self-interest and the fundamental ethical postulate of individual freedom. This creative destruction has its dynamics from the power conflicts in the real sector in terms of real income distribution. The conflicts are reinforced by the financial sector in terms of financial asset effects on the real variables through asset transfers in the game of gambling and risk-taking to shape income distribution. The basic institutional organization of the economic structure and the power distribution in the Schumpeterian socialist political economy is provided in Figure 3.4.

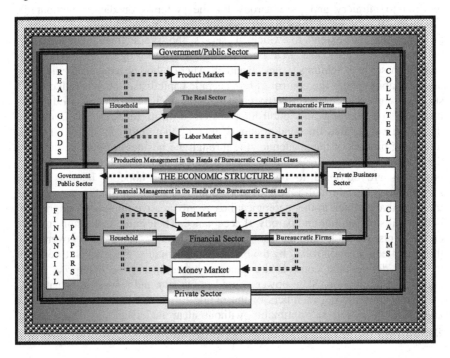

Fig. 3.4 The Economic Space, Aggregative Markets and the Organization of the Real and Financial Sectors of Domestic Bureaucratic Capitalism of a Schumpeterian Socialist Political Economy

In addition to the organization of the structures of the political economy, is the organization of the relative decision-power structures through which decision-choice actions are undertaken in the social setup. The power structure complementing the economic structure to constitute the three building blocks is presented in Figure 3.5 where there is economic power in the economic structure, political power in the political structure and legal power in the legal structure. The relations of the powers and structures constitute the basic organizational structure of the political economy [R4.23][R4.24][R4.56][R4.120].

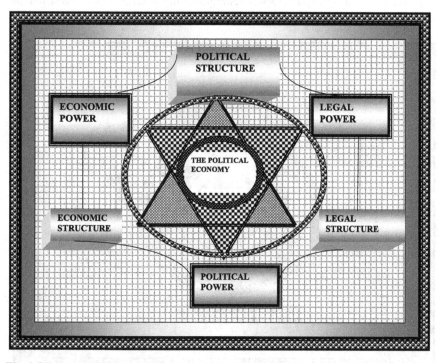

Fig. 3.5 Epistemic Geometry of Relations of Power and Structures of the Political Economy as Power Relations among individual and Collective Decision-Choice Actions in the Structures of Economics, Law and Politics

Given the political economy and the respective powers of the structures, we are then confronted with the problem as to who controls and exercises the social decision-making powers in the various structures. In the Schumpeterian entrepreneurial political economy, economic power is in the hands of the entrepreneurial capitalist class where such power is exercised in private production decisions. The political and legal powers are in the hands of the social decision-making core where the powers are exercised on behave of the general public for fairness, stability and social control for creating an effective market democracy. In the Schumpeterian bureaucratic capitalist political economy, economic power is in the hands of the bureaucratic capitalist class where such power is exercised in private production decisions for profits and self-interests. The political and legal powers which used to be in control of the social decision-making core are now directly or indirectly in the hands of the bureaucratic capitalist class, where the powers are exercised not on behalf of the general public and labor, but on behalf of the capitalist and private business class for rent-seeking, stability and social control to enhance profits and distort the distribution of income. At this level of social transformation, the market democracy is transformed to a market mockery of democracy. This is a Schumpeterian socialist political economy under the control of the bureaucratic capitalist. In a sense, it is socialism of the rich through the control of productive capital. The interactive relations of the structures, power distribution and the markets are shown in Figure 3.6.

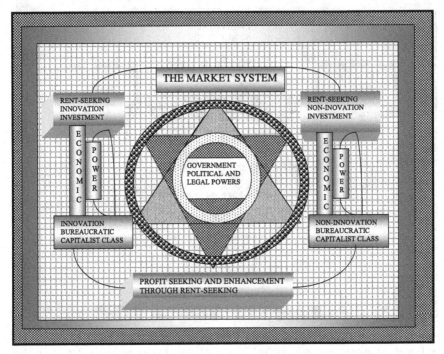

Fig. 3.6 The Cognitive Geometry of the Relational Structures of the State as the Public Sector, the Market System and the Power Distribution in the Schumpeterian Bureaucratic Capitalist Political Economy where the Political, Legal and Economic Powers are controlled by the Bureaucratic Capitalist Class

A few critical observations on the Schumpeterian political economy of the socialist type are important to be brought into focus, relative to the Marxian political economy of the socialist type. Both Schumpeter and Marx define socialism in terms of power relations and the ownership of the means of production. In a Schumpeterian political economy, capitalist socialism is attained when the bureaucratic capitalist class controls the powers of the three sectors for the benefit of the business aristocracy. The power of labor is completely dismantled and reduced to nothingness in the democratic collective decision-choice space where labor is effectively disenfranchised and only playing a subservient role in the market mockery of democracy. In a Marxian political economy, peoples' socialism is attained when labor as a class controls the powers of the three sectors for the benefit of the public. In this respect, a mixed economy may be defined as a political economy where the three powers are distributed between labor and capital. By combining the organizations of the economic structures, the powers and the distribution of the social decision-making powers, we obtain an epistemic look of the Schumpeterian bureaucratic political economy in terms of the exercise of power for the making of laws, rules and regulations and their implementation. The organization and the interactive structures of the regulations are illustrated in Figure 3.7. It is important to note that the Schumpeterian socialist political economy is the same as a bureaucratic capitalist political economy. The dynamics of change has nothing to do with labor which performs a passive role as surrogate capital under the control of the business aristocracy in the Schumpeterian political economy.

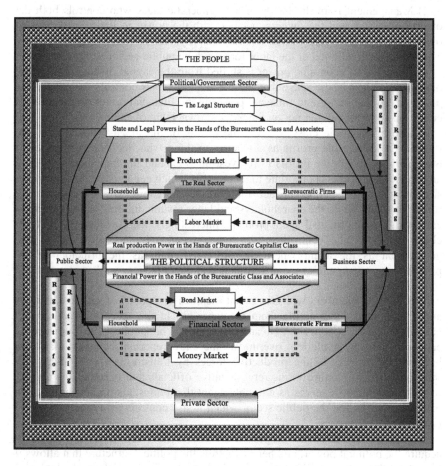

Fig. 3.7 The Economic Space, Aggregative Markets and Decision-Power Distribution for Law Making, Regulations and Controls in the Domestic Bureaucratic Capitalism of a Schumpeterian Political Economy

3.5 The Central Features of the Schumpeterial Socialist Political Economy and Bureaucratic Capitalism

The central features of bureaucratic capitalism and how it relates to rent-seeking and corruption in the democratic decision-choice process may be summarized as:

a) There is a separation between ownership and management.

b) There is some form of democratization of ownership of capital through the financial sector, where portions of ownership become items on constant sale with the ideology of collective ownership called shareholders through the financial markets, and where the shareholders have no decision-making power.

c) Management is in the hands of a technocratic class who controls both the owners of capital and the owners of labor with the power to issue certificates of ownership of capital and power to offer employment to labor, thus effectively controlling the livelihood of the labor force and the public-sector activities.

d) The locus of economic power is shifted from the capital owning class to the bureaucratic class in the sense that the bureaucratic class controls the *production power* in the real sector and financial power in the financial sector on the basis of education and training, but not on the basis of ownership of capital.

e) The bureaucratic class receives the profits of the mega-firms and has the power to distribute the profits as dividends to those who hold the certificates of ownership, or to use the profits to reinvest or to expand the bureaucratic firm in any direction it deems appropriate to its interests which need not coincide with the shareholders or the public. In fact, the bureaucratic capitalist class assumes the role as an agent while the shareholders and labor assume the role as the principal, creating a classic principal-agent problem. Interestingly, the bureaucratic capitalist class, by indirectly controlling the government, assumes the role of the government as the agent where the citizens remain as the principal in the principal-agent structure of the government and the citizen, as well as play the role of agent to itself.

f) It pays its members with exorbitant incomes composed of salaries, bonuses and capital shares that entitle them to ownership in the process in addition to handsome fringe benefits which are classified to be public. The bureaucratic capitalist class, by evolution, overthrows the entrepreneurial capitalist class and becomes the owners of the productive forces by transformation. Similarly, by transformation it overthrows the representative officials and becomes the owners of the government. It serves as a double agent that represents the business sector which it controls and the public sector which it also controls by a remote political process.

g) With the use of economic power, the bureaucratic capitalist class comes to acquire the political power, and hence controls the political structure that allows it to define the national interests and social goal-objective set in the political structure, as well as assert the directions of laws, rules and regulations in the legal structure in the favor of the bureaucratic class, thus creating political oligarchy without regard to the masses. Effectively, but indirectly, it overthrows the public and becomes the owners of the political and legal powers. Voting in the democratic collective decision-choice process in the Schumpeterian political economy is simply an exercise in deception for pacification and rubber-stamping under the ideology of government by the people, of the people and for the people. The real truth in the Schumpeterian socialism there is a government to deceive the people by means of voter-participation, where the government is of the bureaucratic class for the bureaucratic class. The voter-participation merely recycles different members from the same bureaucratic class into the social decision-making core that constitutes the governing class to deliver governance in favor of the members of the bureaucratic class.

   h) With direct or indirect control of the power of the State, the bureaucratic class has succeeded in creating bureaucratic socialism not for the masses but for the bureaucratic capitalist class, where profits are privatized and losses are socialized through some complicated incentive and tax system where the business sector is knit by horizontal and vertical integrations as well as in bed with the government.

   i) It controls the social communication channels as part of bureaucratic capitalism and uses them to promote ideology of people's capitalism, patriotism and militarism in support of its interest. It follows management practices based on scientific rationality with systematic decision process and planning that enhance efficiency, increase productivity and national wealth as well as increase economic inequality through the principle of capital shareholding transfers and external management process. With the effective control of business schools in the universities and the sources of research founding in the knowledge areas of the university essential to its operations and profit seeking, it succeeds to promote this ideology as a foundation of progressive civilization by controlling both the state of knowledge production and the knowledge production enterprise. Research in science, technology, engineering and mathematics are supported to the extent to which the results have actual and potential outcomes to enhance the progress of the interests of the bureaucratic class. Philanthropic activities are created as ideological cover and to reduce tax obligations.

   The basic characteristics that have been outlined here are present in the major industrialized economies that claim to function under the principles of a democratic decision-choice system. They present a troubling vision of global tomorrow where labor is subservient, playing the role of surrogate capital, and labor organizations are slowly and systematically dismantled leading to ineffective organized-labor resistance. In a sense, labor is moved to a different level on the leader of slavery. Consumer protection is reduced to nothingness by the rules and actions in the legal structure. The troubling vision is that the increasing concentration of power in the hands of the bureaucratic capitalist class is an important gathering threat and growing menace to a true global democratic collective decision-choice system based on individual freedom within the collective freedom, and the practice of individual interest within the collective interest that will allow the construction of an appropriate institutional configuration for a true democratic collective decision-choice system that will support freedom and justice. It also presents a danger to the foundations of the social fabric and work ethics on the basis of wealth-producing risk-taking in the real sector.

With the emergence of bureaucratic capital, the industrial entrepreneurs with innovation, innovation investment and wealth-producing risk-taking have become financial entrepreneurs with financial innovation, financial-innovation investment in the production of paper assets and non-wealth-producing risk-taking elements. Within bureaucratic capitalism, real wealth has been replaced by phantom wealth through a process of financial engineering of paper assets that are presented as accounting documents without real output support. The social problem here, is that there is a growing menace that personal wealth-creation takes place in the financial markets where one moves paper money to create paper wealth, and this is ideologically integrated into the national psychic that paper assets and nominal accounts constitute the real wealth of a nation. In this respect, there arises a social problem regarding the notion that financial accounts and nominal money of a nation without comparable support of real goods and services is nothing but phantom wealth. In a closed political economy this has a tendency to generate inflation and pauperize the masses. In an open-political economy this has a tendency to generate external debt through imports to support real consumption and real investment.

3.6 Oligarchic Socialism, Phantom-Wealth Creation and Rent-Seeking in the Real and Financial Sectors

Let us examine the relational process of rent-seeking and phantom-wealth creation under oligarchic socialism as a transformation from bureaucratic capitalism in the sense of Schumpeter [R4.100] [R4.101]. Rent-seeking is neither sector-dependent nor sector-specific. It can take place in either the real sector or financial sector or both as the government creates and repeals laws, rules and regulations to affect decision-choice actions in the real-sector markets and financial-sector markets. The point here is that government laws and regulations have the potential to create rent-seeking environments in the market system operated under personal interest and profit seeking. The rent-seeking activities are undertaken to enhance individual profits and benefits. The rent-seeking environment may be divided into a real sector rent-seeking environment and a financial sector rent-seeking environment depending on the makers and enforcers of the laws and regulations. Rent-seeking does not arise in a capitalist political economy without government to manage the institutions of the market.

In general, the activities in the real sector produce real wealth, and it is under this factor that speculative activities of the entrepreneurs create individual and national real wealth, measured in some aggregate index of goods and services. The ultimate driving force is the use of profit to enhance profit through innovative and non-innovative investment processes by entrepreneurs. The rules and regulations are primarily created by the social decision-making core for managing fairness of competition, discouraging undue advantages of monopolies and others, and protection of the public and consumers from industrial misdeeds such as product misrepresentation and others. The rules and regulations in the market also involve managing protection for quality and safety and against dangerous methods of production and others, in addition to preventing the exploitive environment of the labor force by overzealous profiteers. These rules and regulations positively affect costs and negatively affect profits to enhance qualitative characteristics of the social setup. Here, a rent-seeking environment translates into cost-avoidance environment to enhance profits. Rent-creation translates into exploiting loopholes in the laws and regulations for cost avoidance, where rent-preservation translates into exploiting weaknesses in enforcement and rent harvesting translates into taking advantage of weak or lack of enforcement.

Generally, however, the social decision-making core regulates for quality, safety, fairness in competition and pricing in the real sector. The basic goal of the government in the entrepreneurial political economy is to help in creating and maintaining reasonably good and robust competitive capitalist markets in the real sector. In this respect, regulation must be flexible, dynamic and yet effective. The legal sector of the regulatory regime, as seen in entrepreneurial capitalism, constrains the goal of bureaucratic capitalism in restraining and reducing competition. The bureaucratic capitalist class in the real sector seeks to control the government to reduce rules, regulations and laws in the legal sector that affect its members' costs as broadly defined at the expense of the public sector. It does this by influencing the decision-making core to reduce both intensive and extensive regulations. It works simultaneously to tighten the laws and regulations against the workforce and consumers while it works to relax the regulatory regime against its members' activities. The rent-seeking in this respect is cost-transferring as well as real wealth-producing but at the expense of the public. The cost transferring reveals itself as bailouts of bureaucratic firms when losses are made, and is supported by profit privatization into basically the personal riches of the members of the bureaucratic class. The rent-seeking in the real sector includes environmental regulation and land-tenure questions.

When we consider the financial sector, things are different as seen from the regulatory process in the legal structure. First we must ask the question: what activities in the financial sector are candidates for the government to regulate from the legal structure? This question is important since the financial sector involves possibilities of creating overlapping financial instruments whose role in the national economy may be vague and misunderstood in terms of the role of the financial sector in supporting the real sector's production-consumption activities. The financial instruments, as financial goods and services, are debt instruments as well as risk-avoidance and risk-transfer instruments. Some of these financial instruments may have some commodity support from the real sector (collateral assets) and some may have no commodity support in the real sector. The conditions and technology for their creation are different from the conditions and technology for the creation of real goods and services. Nonetheless, the decisions in the two sectors must be coordinated to create and maintain strict dynamic production-consumption equilibrium and stability which we shall refer to as strict *real-financial equilibrium* in the full dynamic sense. This dynamic stability in the economic structure is required for sustainability of the general dynamic stability of the political economy involving the political and legal structures. Because of the differences in the production activities in the real and financial sectors, there are important differences in the nature of rent-seeking in both sectors. We shall explore these differences and their relational structures as they relate to the environment of phantom-wealth creation.

3.6.1 A Structure of Rent-Seeking in the Financial Sector

To understand the problem of rent seeking in the financial sector, we need to construct a simple but reasonable relationship between the real sector and the financial sector. At the level of the current account of the wealth-financial claim balance sheet, the financial sector produces financial claims over commodities and services but not commodities themselves, while the real commodities and services serve as collateral or guarantor of the wealth of the financial instruments. These financial claims appear as coupons and have no value without the production of real goods and services in the real sector. The management and the activities that enter into the production of the financial claims will enter into the real sector as goods and services. The wealth of the nation is produced in the real sector as goods and services. For every entry in the financial sector, there is a corresponding entry in the real sector that produces a balance sheet. A simple and consolidated structure of wealth-financial claims is shown in Table 3.1.

The coupons in the money market are direct legal tenders with fixed values and cannot be bought and sold in the market. The coupons in the debt market are indirect legal tenders with varying values that can be sold and bought at varying prices in the bond market. The financial claims are useless to the holder if they have no commodity support which may be referred to as the *real collateral* of the national financial papers. All the real commodities are produced in the real sector whose support produces the social trust to hold financial instruments in the debt market, and money in the money market. From the wealth-financial claims balance sheet, we obtain the total consumer goods, $\mathbb{Q}_T^C$ available for financial claims in the market, as the sum of consumer inventories $\mathbb{V}_C$, and currently produced consumer goods $\mathbb{Q}_C$. The total available consumer goods for financial claims is $\mathbb{Q}_T^C = \left(\mathbb{Q}_C + \mathbb{V}_C \right)$. The total investment goods in the market $\mathbb{Q}_T^I$ for financial claims, is the sum of investment goods inventories $\mathbb{V}_I$ and currently produced investment goods $\mathbb{Q}_I$. The total available investment goods for financial claims is $\mathbb{Q}_T^I = \left(\mathbb{Q}_I + \mathbb{V}_I \right)$. The total housing-investment goods in the market, $\mathbb{Q}_T^H$ for financial claims is the sum of housing-investment goods inventories $\mathbb{V}_H$ and currently produced housing-investment goods $\mathbb{Q}_H$. The total available housing for financial claims equals $\mathbb{Q}_T^H = \left(\mathbb{Q}_H + \mathbb{V}_H \right)$. The total goods in the market $\mathbb{Q}_T$ for financial claims at any time is the composite sum

Table 3.1 A Consolated National Wealth-And-Financial Claims Balance Sheet

TOTAL REAL SECTOR-ASSET COLLATERAL SUPPORT FOR FINANCIAL CLAIMS (LIABILITIES IN THE FINANCIAL SECTOR)	TOTAL FINANCIAL SECTOR LIABILITIES OR CLAIMS OVER NATIONAL WEALTH (COMMODITIES AND ASSETS)
COMMODITY MARKET	A MONEY MARKET
A CURRENT PRODUCT ACCOUNT	CURRENT CASH-IN HAND 1) Household .... M_H
1. Current Consumption Goods Produced, Q_C	2) Firms................. M_P
2. Current Investment Goods Produced Q_I	3) Banks .... M_B
3. Current Housing investment Produced Q_H	4) Lending (Savings and loan)
4. Current Social Investment Produced Q_S	Institutions...... M_L
5. Current Consumer goods Inventories V_C	
6. Current Investment goods inventories V_I	5) Government................M_G
7. Housing inventories Investment V_H	
	B DEBT (Loanable Funds) MARKET FINANCIAL PAPERS
B CURRENT CAPITAL ACCOUNT	6) Private Bonds..................$\mathbb{B}_P$
8. Accumulated Productive Capital	7) Government Bonds............$\mathbb{B}_G$
of the Private Sector .... $\mathbb{K}_P$	8) Consumer Loans................$\mathbb{L}_C$
9. Accumulated Social Infra-structure Capital $\mathbb{K}_S$	9) Firms' loans.....................$\mathbb{L}_F$
10. Accumulated consumer Durables...........$\mathbb{K}_D$	10) Other Lending Institutions... $\mathbb{L}_O$
11. Accumulated Domestic Housing	11) Consumer Credit.............$\mathbb{C}_C$
Investment $\mathbb{K}_H$	12) Commercial Credit..........$\mathbb{C}_F$
12. Accumulated Service Capacity Investment $\mathbb{K}_A$	13) Insurance Papers ....$\mathbb{C}_I$
13. Accumulated Depreciation Investment .... $\mathbb{K}_\delta$	CAPITAL GAINS

$\mathbb{Q}_T = \left(\mathbb{Q}_T^C + \mathbb{Q}_T^I + \mathbb{Q}_T^H \right)$, all valued in the current or selected base year aggregate price index of real commodities. In other words, the total real goods and services for current financial claims is the sum of consumer goods, investment goods and housing in the market. The sum of these commodities constitute the total collateral for the current financial claims at any current time. It is these commodities that give value and incentives to hold financial papers either as current or future claims. In every moment of time the current financial claims must be distinguished from the previous financial claims that define ownerships of previous real national wealth. The current financial claims define the distribution of the possible ownership of the current national wealth. The used past financial claims define the distribution of ownerships of the existing national wealth. The unused financial claims in the political economy define the possible distribution of future ownerships of the real national wealth. It is this future possibility of ownership of real national wealth that creates the incentive to hold different forms of financial papers.

At the level of accumulated wealth, the capital account is such that the total current productive capital of the private sector $\mathbb{K}_T^P$ is the sum of accumulated private productive capital $\mathbb{K}_P$ and currently produced investment goods $\mathbb{Q}_I$, and investment-goods inventories $\mathbb{V}_I$ thus we have $\mathbb{K}_T^I = \left(\mathbb{K}_P + \mathbb{Q}_I + \mathbb{V}_I \right)$. The total domestic housing capital $\mathbb{K}_T^H$ is the sum of accumulated domestic housing investment (or capital) $\mathbb{K}_H$ and currently produced housing-investment goods $\mathbb{Q}_H$ and housing inventories $\mathbb{V}_H$ thus $\mathbb{K}_T^H = \left(\mathbb{K}_H + \mathbb{Q}_H + \mathbb{V}_H \right)$. The total social infra-structure capital $\mathbb{K}_T^S$ is the sum of accumulated social infra-structure investment $\mathbb{K}_S$ and current social investment produced $\mathbb{Q}_S$, and hence $\mathbb{K}_T^S = \left(\mathbb{K}_S + \mathbb{Q}_S \right)$ since there are no infrastructure inventories. The social infrastructure capital and investment are composed of defense and non-defense infrastructure components. In addition to these aggregates, we have accumulated consumer durables $\mathbb{K}_D$, accumulated service capacity investment, $\mathbb{K}_A$ and accumulated depreciation investment $\mathbb{K}_\delta$. These items are the real wealth of a nation that belongs to different members or groups in the nation by the process of income distribution. The total additions of the national real wealth $\mathbb{W}$ in the market at any-time is the sum of available consumer goods in the market $\mathbb{Q}_T^C$, current available investment goods in the market $\mathbb{Q}_T^I$ and current available housing goods in the market $\mathbb{Q}_T^H$ hence the total goods in the market $\mathbb{Q}_T = \left(\mathbb{Q}_T^C + \mathbb{Q}_T^I + \mathbb{Q}_T^H \right)$. The accumulated productive capital $\mathbb{K}_P$ and the accumulated housing capital $\mathbb{K}_H$, infrastructure investment $\mathbb{K}_T^S$, accumulated consumer durables $\mathbb{K}_D$ and the accumulated service capacity investment $\mathbb{K}_A$ are holds for financial claims. The government buildings, public institutional investment, and others are included in the total social investment. The total wealth of the nation in a closed economy may be written as structured in Figure 3.8. The wealth-income accounting and distribution are such that the total domestic production of goods and services is divided between the private and public sectors as claims. The total of claims, on goods and services and wealth, is divided among workers, capitalists and the joint-claim held by the government that represents the public interests for collective ownership. The national wealth is composed of accumulated wealth and its growth.

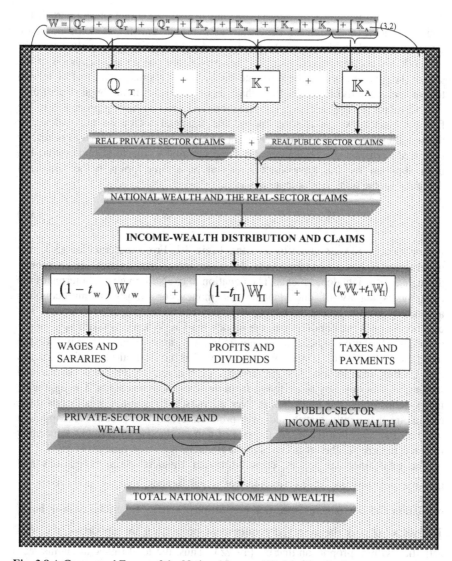

Fig. 3.8 A Conceptual Frame of the National Income-Wealth Distribution

The distribution of the claims shows the structure of wealth-income distribution. Let the distribution of wealth and income be between profits and the assets held by profit earners and wage income and assets held by wage-income earners such that $\mathbb{W} = (1 - t_w) \mathbb{W}_w + (1 - t_\Pi) \mathbb{W}_\Pi + (t_w \mathbb{W}_w + t_\Pi \mathbb{W}_\Pi)$ where the values t_w and t_Π are tax rates on wages and profits. The claims are held as coupons and financial instruments over the national wealth which may be divided into private and public sector claims. In a closed economy a sale of any part of the national wealth is simply an internal transfer of claims. A sale of any part of the national wealth to foreign entities is an international transfer of domestic claim over the domestic real wealth. Privatization of public entities is simply a transfer of public claims into private claims. Thus privatization is an augmentation of the size of the private claims. These private-public sector claims are intimately related to rent-seeking activities under the behaviors of the financial and legal structures.

To explain the rent-seeking, rent-creation, and rent-harvesting processes in the financial sector, we shall superimpose the financial sector on the real sector through the commodity claims. Let us suppose that the financial system is strictly regulated such that the total money-in-hand is equal to the total money that has been supplied by the government, in that at a time point before financial deregulation the total available money is:

$$\mathbb{M}_{G_0} = \left(\mathbb{M}_{H_0} + \mathbb{M}_{F_0} + \mathbb{M}_{B_0} + \mathbb{M}_{L_0} \right) \tag{3.3}$$

The values $\mathbb{M}$'s are as defined in the Table 3.1. From the equation (3.3), the government holds no money since the government can always create money through appropriate departments. The equation may also be viewed in terms of the economist's supply of and demand for money, or simply as an economic accounting identity where the supply of money equals the demand for money. The equation defines the conditions in the money market. When it is taken as a supply and demand relation rather than economic identity, an economic logic is called in to establish theories of demand for money by households, real-sector firms, banks and lending institutions. The preferences of the real-sector firms, banks and lending institutions may be different from those of the households. The aggregate money constitutes the total claims over all outstanding initial wealth but these claims are differently distributed over productive and non-productive capital. It will become clear that the larger size of non-productive investment, especially military, the greater the instabilities in the financial sector as time evolves.

The total financial papers, constituting net financial-claim obligations and tradable instruments, $\mathbb{D}_{T_0}$ as an indirect legal tender that is authorized by regulation and law is some specified proportion $(1-\alpha)$ of the total cash-in-hand. The value α is the reserved requirement proportion in support of the loan activities with a proportion that is $(1-\alpha)$. Thus, the total financial claims $\mathbb{F}_T$, over the total real-sector collateral is specified in the initial time as:

$$\mathbb{F}_{T_0} = (1+\alpha)\mathbb{M}_{G_0} = \mathbb{M}_{G_0} + \left(\alpha_1 \mathbb{B}_P + \alpha_2 \mathbb{B}_G + \alpha_3 \mathbb{L}_C + \alpha_4 \mathbb{L}_F + \alpha_5 \mathbb{L}_O + \alpha_6 \mathbb{C}_C + \alpha_7 \mathbb{C}_F + \alpha_8 \mathbb{C}_I \right) \tag{3.4}$$

The value $\alpha = \sum_{i=1}^{8} \alpha_i$ is the financial claims' distribution parameters. The value of the regulatory parameter (α) has a multiplier effect in the financial system which may be approximated by $\left(\frac{1}{\alpha}\right) = \left(1 \middle/ \sum_{i=1}^{8} \alpha_i\right)$. If the government's money

creation always covers the value of output produced at each point of time, then the

parameter, $\alpha = \sum_{i=1}^{8} \alpha_i$, is a measure of financial security and $(1-\alpha)$ is a measure

of the *systemic risk*, in that $(1-\alpha) \mathbb{M}_{G_0}$ has no real commodity support. The

systemic risk is managed by the decision-making core by regulating $\alpha = \sum_{i=1}^{8} \alpha_i$ in

order to regulate the behavior of the systemic risk in the political economy from the debt market. As α is increased the systemic risk falls but rises as α falls. As it relates to the political economy, the volume of $(1-\alpha) \mathbb{M}_{G_0}$ is basically a

transfer of real-sector claims from those who own the accumulated wealth or inventories to those who are deficit in real commodities. In this way, the surplus and the deficit spenders are brought together by the behavior of the debt market which generates financial papers as instruments of financing and storing obligations. At the macroeconomic level, the economic analysis becomes more complex with increasing difficulty as we seek to relate total national savings to accumulated capital stock, general durables and inventories of all kinds to the total financial obligations and claims, as well as examine the policy interactions of political, legal and economic variables.

From the real sector, and at the level of the capital stock account, it must be the case that the national stock of productive capital plus accumulated consumer durables plus accumulated social infrastructure capital must be equal to accumulated savings. The total current real savings must be equal to the total current inventories plus the currently finished infrastructure constructions. From the financial sector, and at the level of the money market, if the supply of money by the government from a specified initial period is equal to the value of the goods and services at any time period, then the total money in circulation must be approximately equal to the value of the national stock of productive capital, plus accumulated consumer durables plus accumulated social infrastructure capital plus total inventories at certain base year prices.

In this way, an effective governmental regulation to maintain the system's stability will require regulating α and its distribution over the various claims within acceptable bounds of the system's strength. The debt market may be seen as an institution to lubricate the economy through reconciliation between deficit and surplus spenders for a fee, and with regard to a wide range of risky conditions for the individual business and non-business entities in the political economy. The fee is the cost of services including the cost of waiting. The nature of the distribution of the value of $(1-\alpha)$ over the institutions of loanable funds will

determine the degree of overall systemic risk generated by the financial sector. It can also degenerate to a place of gambling when regulations are weakened through lack of enforcement or dismantlement through the legal structure. It is this aspect of production in the financial market which the types of the financial products must be regulated. In other words, the governmental regulation must be both qualitative and quantitative to manage the systemic risk. Quantitative regulation refers to changes in the space of quantity production of total financial instruments in the system. Qualitative regulation refers to the number of types of financial instruments that can be produced by the financial sector and by what financial industries.

Ideally, the strict real-financial equilibrium requires that the value of produced goods and services in the real sector should be equal to the value of the created financial assets in the financial sector valued at a given unit of account at each time point. The total financial assets from the financial sector, composed of money and financial instruments, is taken here as the total purchasing power of the economy, in that each unit of the purchasing power has a claim over some commodities that have been produced in the real sector from an initialized period to the present whether such claims are direct or indirect legal tenders. The total value of the financial papers constitutes a liability to the real sector. The total value of the sum of goods and services produced from an initialized period to the present or any time period less depreciation and spoilage in the real sector constitutes the wealth of the nation for any present time. The total value in each year will constitute a yearly increase in the wealth of the nation while the total value financial instruments in the year will constitute the yearly increase in claims. The yearly increase in the real national wealth constitutes the yearly collateral for the yearly increase in the financial claims. Given the existing quantity of money, a question may then be raised as to what are the bases of creating financial assets. The case of strict real-financial equilibrium requires that every financial asset must be collateralized from the real sector. This, of course is not the case.

The concept of depreciation of the national wealth needs some explanation as it relates to the national wealth-finance balances. The depreciation of the national wealth includes depreciation of productive capital, depreciation of social infrastructure, nationally perished and used commodities for continuous usage such as food, water, electricity, gas and other service input to life. The depreciation of the productive capital includes the productive assets that have been withdrawn from production plus the depreciation of assets that are in production. All these will constitute the accumulated depreciation as the total depreciation of the national wealth. The role of such depreciation in the calculus of thought is that the total national financial assets will be greater than the national wealth to the extent to which inflation factors are accounted for. The excess of these financial assets will have claims over real commodities that are technically not in existence and hence must compete against the existing claims of the national wealth. For the general production-consumption duality, the money corresponding to one-period production of goods and services becomes the financial wealth. The collateral for those debts that have no collateral is the future production of real goods and services in the political economy.

If the financial sector ceases to be an institution for reconciling the disparity between deficit and surplus spenders but generates sub-markets within itself on the basis of risk-income profile, then the financial papers in the debt market become financial commodities that are tradable not in the traditional sense of mobilizing savings in support of investment in the savings-investment equality constraint on the economic growth-development process. The commodity papers have the characteristics of risk, returns and interest rate that are marketed for a price where such a price varies in the market for the same financial commodity. The financial market becomes segmented into a current market with a current price, a forward market with a forward price and a future market with a future price.

Furthermore, the commoditization of the financial instrument carries to the market conditions of quality segmentation by the characteristic of the risk-return profiles that help to determine the price of a financial commodity. The disparity between either current and future prices or forward and future prices establishes the concepts of capital gains and losses which become the motivation for gambling in the debt market for whatever the money market is. The distribution of the disparities under a differential distributions of uncertainty-risk states of nature is the reward distribution in this socially approved gambling. The introduction of a risk-return profile of each financial commodity allows the debt institutions to create different commodities in the debt market where quality is defined by grading, just as different commodities are created in the markets of the real sector with different grading systems. The trading is done in terms of present, forward prices for capital gains or losses in the ruling future price. The production of financial papers and the buying and selling of financial papers for capital gains have transformed debt institutions in support of real sector productions into gambling institutions, where transfers of claims become the transfers of debts with varying degrees of risk for reward.

Under such conditions, the debt market has been transformed from the role of reconciling deficit and surplus spenders for investment, production and consumption activities in the real sector into wealth-savings transfers and over-consumption without real-wealth support. Disposable income is augmented by debt creation and consumption is amplified without corresponding production. To maintain the wealth-savings transfers and over-consumption as financial output activities, there are continual product differentiations of financial papers in terms of risk-return profiles and maturity periods by the financial firms. The production of these commodity financial papers may be done and usually done independently and without regard to the real sector production even though they have claims over real commodities and national wealth. In this respect, the growth of financial wealth may have no collateral in the real sector in terms of real wealth support. The result is the creation of *phantom wealth* that exists as financial papers leading to the emergence of an increasing economic bubble between the real sector and the financial sector where the acceleration of the growth of the total national financial assets increasingly exceeds the acceleration of the growth of real national wealth. The phantom wealth is the difference between the total financial assets and total national wealth valued in terms of a given unit of account. The economic bubble, at any time point, is the difference between the speed of growth of the total financial assets and that of the growth of the real national wealth. The phantom wealth is translated as the total financial assets that have no real wealth support. A creation of financial paper is a creation of claim over real wealth and hence a transfer of real wealth to be paid at some specified date. The reserve requirement ratio provides the maximum size of the phantom worth that the financial sector can create.

The greater the growth of real output (wealth) and the slower the growth of financial wealth, the smaller is the phantom wealth and hence the smaller the economic bubble. Alternatively, the greater the growth of financial wealth and the slower the growth of real wealth, the greater is the phantom wealth and hence the greater is the economic bubble between the two sectors. Let us keep in mind that the total output in the real sector is the collateral for the total output in the financial sector which is made up of direct legal tender in the money market and indirect legal tenders in the debt market. The only constraints on the production and selling of financial papers are the government's extensive and intensive regulations relative to governmental monetary activities and real sector productive activities. The less intensive and extensive the regulations are, the greater is the possibility for money-making by others through increasing financial engineering for empty financial wealth for some.

The role of the financial institutions as parts of bureaucratic oligarchy is to work through the political structure that they indirectly or directly control to effect intensive and extensive regulations in the debt market in order to create opportunities for rent-seeking activities in terms of possibilities of rent-creation and probability of rent-harvesting. The activities and engineering innovation in the debt market is to disconnect the financial sector decisions completely from the real sector information for money game where people move money to make money in order to create financial wealth without real production that creates real national wealth. The financial regulatory process involves changes in the value of the reserve requirement ratio for quantitative control, the structural institutions that can produce financial papers of indirect legal claims and the nature of the financial output and its distribution over risk for qualitative control. The weakening of the regulation can relax the quantitative and qualitative constraints on the institutions and incorporate new ones to widen the space of the financial production to expand the space of the *phantom wealth*, the *bubble* and the *systemic risk*.

3.7 A Note on Rent-Seeking in the Marxian Political Economy

There is little to say about rent-seeking in the Marxian economy. To see why this is the case let, us outline the basic characteristics of the Marxian political economy and the Marxian social transformation dynamics.

1. The primary factor that determines the wealth of nations is labor and hence national wealth and income are reducible to labor units.

2. The primary relations in the society are economic, and all other relations and policies are derived from these primary relations.

3. The primary factor that determines economic relations, social decision-making power and hence the distribution of social income is the control of ownership of productive capital in order to control the economic power.

4. The primary value system that determines the framework of social decision-making regarding production, politics and law is that of the owners of capital, which runs counter to that of workers who produce national wealth and income with their labor.

5. The conflicts between the value systems of capitalists and workers regarding production, distribution, application and enforcement of rules and regulation lead to intense struggles between the two groups to control the institutions of government that have been set up to organize the society. These struggles provide the explanation for the drive to control the state apparatus by either labor as a class or the capitalists as a class.

6. The primary mechanism for socio-political transformation is the conflicts between workers and capitalists at the arena of production, distribution and the control of the legal structure through the political structure.

7. The struggle over equitable distribution of real income and fairness in the legal and political structures is the driving force in socio-political decisions that affect economic actions.

8. Labor is the primary factor for creating capital and technology, after which it becomes controlled by capital through its ownership, as well as becomes displaced out of work by a process of capital-labor substitution technology, which further creates capital-labor conflict in the capital-labor duality.

9. The capital-labor conflict is expressed in the general decision-choice space composed of economic structure, political structure and legal structure. Here the working class, reasoning from the perception of its interest, seeks to control the government or the decision-making core to change the legal structure in its favor. Similarly, the capitalist class, reasoning from its profit-motive interests, seeks to control the government to change the legal structure in its favor.

10. The resultant dynamics in the Marxian political economy is where socialism is established by labor and for labor in the sense of workers socialism, where the workers control the government, control the political structure thus acquiring the power to make laws to change the legal structure by abolishing private capital ownership and profit motive of production, in favor of state ownership of capital and public interest in production.

11. The end result is that the workers come to control the economic power, the political power and the legal power in the service of the general public.

The state takeover of the production and distribution mechanisms in the economic structure effectively does away with the possibility of rent-seeking activities by private production on the basis of profit. The process of development of indirect controls from the private sector to create influence and change the legal structure is also done away with. The conditions of private financial-real sector risk in decision-making in the political economy are done away with and hence there is no market for individual risk-taking on the basis of which insurance enterprises arise. Conditions of private lending and borrowing as economic activities are dismantled. The social decision-making core (government) has power concentration that allows it to control the political, economic and legal structures. The problem that arises here is the emergence of the possibility of governing bureaucratic corruption and violation of the public trust that comes with governance by the social decision-making core. The space of the governing bureaucratic corruption and oppression has a tendency to widen and increase in intensity with political incumbency. It is the existence of this incumbency possibility that provides a justification for the government to be served by a continuous succession of different social decision-making core which is created by an electoral process with the hope of bringing in different governance with reduced corruption. It is also this electoral process that forms an important social calculus for resolving conflicts in the individual preferences in the democratic collective decision-choice system without violence.

Chapter 4
Fuzzy-Stochastic Information, Financial Bubbles, Systemic Risk, Creative Destruction and Rent-Seeking Society in the Schumpeterian Political Economy

The previous chapter ended up with the nature of a money-managed economy and the financial-paper game. To follow and appreciate the complexity of the financial-paper game and the corresponding rent-seeking activities, we must keep in mind that the political economy is such that the two sectors are connected to define the production-consumption process. The connection reveals itself in the framework where the financial sector's output goes to support the real income distribution and lubricate the exchange flows of goods and services. Additionally, it provides the service of real-value storage for real-value claims through the creation of financial assets. It also provides a mode of wealth-transfers and increasing individual or group purchasing power through credit or debt creation in reconciling conflicts between the deficit and surplus spenders in the real sector. Debt creation and buying and selling of debt are simply continual transfers of real claims over commodities and services from surplus-spenders to deficit-spenders and vice versa through a complicated system of creation of financial assets or different forms of IOUs or financial instruments and complex institutions of the debt market and sub-markets for marketing different grades of risk-income profiles. The financial instruments are the output of the bond market augmented by credit and loans which together may be correctly called the *debt market*, where the political economy is transformed into the *debt-managed political economy* and the total market is transformed into *a market of games and gaming*.

The information structure required for tactical and strategic decision-making becomes more complex. Games are not about collective welfare improvement or social progress; they are about individual welfare improvement. It is here that understanding the information structure becomes crucial for the aggregate outcome of the events in the political economy. The complex information structure is related to uncertainty which generates risks. The complex information structure is called *fuzzy-stochastic information* which is composed of defective and deceptive substructures. The *defective information substructure* is made up of

K.K. Dompere, *Fuzziness, Democracy, Control and Collective Decision-Choice System*, Studies in Systems, Decision and Control 5, DOI: 10.1007/978-3-319-05329-5_4, © Springer International Publishing Switzerland 2014 fuzzy characteristics and stochastic characteristics which are called *fuzzy information substructure* and *stochastic information substructure*. The deceptive information substructure is in turn composed of *disinformation substructure* and *misinformation substructure*. The fuzzy information substructure leads to fuzzy uncertainty that generates fuzzy risk. The stochastic information substructure leads to stochastic uncertainty that generates stochastic risk. The fuzzy-stochastic information generates fuzzy-stochastic uncertainty which then produces fuzzy-stochastic risk or stochastic-fuzzy risk. The deceptive information structure generates deceptive uncertainty and associated risk that go to amplify the fuzzy-stochastic risk in the aggregate activities in the political economy. It is here that asymmetric information becomes the central factor in the analysis of behavior in the principal-agent duality. The relational elements of information in the politico-economic games are presented in Figure 4.1.

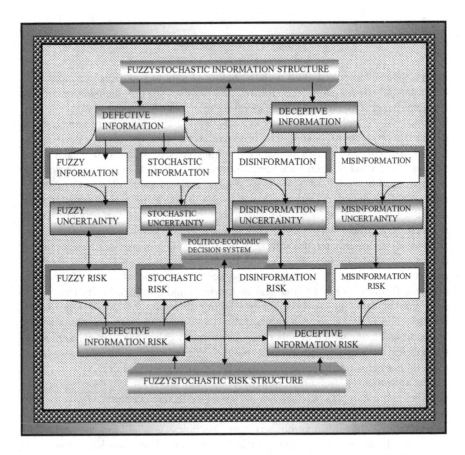

Fig. 4.1 An Epistemic Geometry of the Relational Elements of Types of Information Structures, Uncertainty, and Risk in Decision-choice Systems

By affecting the decision-making process, the general information structure, also establishes the nature of the equilibrating relationship between the real and financial sectors and the nature of the financial bubble. The real sector and the financial sector are in a simultaneous equilibrium when there is no financial bubble. The real-financial equilibrium means that every unit of a debt instrument or a financial asset has equivalent value in the real sector as collateral. Each unit output of the debt market has a price, coupon rate and risk. Each unit of real value in the real sector has a price that allows the calculation of equivalences in both the financial and the real sectors. *Real-financial* disequilibrium exists when there is the presence of a bubble whether such a bubble comes from one or more sub-markets of the aggregate debt market. In other words, there are some financial assets whose values have no collateral support in the real sector for a given time and period. This is equivalent to the statement that there is a disparity between the real sector and the financial sector values at a given unit of account. The disequilibrium between the financial and real sectors creates economic imbalance that may cause an aggregate instability leading to chaotic conditions and a system's failure. The intensity of the system's failure depends on the length of the disequilibrium distance or the size of the bubble as seen from a measure of stability between the two sectors in the political economy.

Every disequilibrium state in the real-financial configuration has a bubble which entails a risk for the political economy. The nature of the equilibrium is essential for the smoothing out consumption-production process, investment-growth process, income-distribution process and financial-savings-real-inventory processes. The financial-savings process takes place in the markets within the financial sector. The real-inventory process takes place in the markets within the real sector. The two sectors interact to bring about temporary equilibrium for any given information set. The size of the risk increases as the size of the bubble increases. In general, there is a sustainable size of the bubble that the real sector is capable of supporting to keep the system stable as well as to allow tradability of individual and submarket risk-income profiles for sustainable systemic risk. The sustainable systemic risk must be in line with financial activities in the support of savings-investment and surplus-deficit consumption processes in the real sector as well as for real wealth generating and surplus-deficit consumption processes for maintaining inventory-production stabilities in the political economy. This sustainable bubble is the largest bubble that the system can sustain to maintain its systemic risk as socially engendered without a failure. There is also a minimum bubble where the system becomes sluggish in the consumption-production-employment configuration due to the nature of income distribution. The maximum sustainable and minimum sustainable bubbles must be related to information and uncertainty in the two sectors. The systemic risk, used in this discussion, takes into account the interactive nature of the real and financial sectors where there is *fuzzy-probability distribution* that real output (wealth) growth will not be able to support the financial bubble due to the unregulated growth of financial wealth. The computable systemic risk is the result of fuzzy-stochastic uncertainty and its value must be computed from the fuzzy-stochastic space that holds the conditions of fuzziness (quality) and limitationality (quantity) of information[R15.4] [R15.5]

[R7.14] [R7.15] [R7.16][R15.13][R15.14]. The financial sector is characterized by financial asset-creation under risk. Such financial asset-creation may have no direct relation to the real-sector production where real goods, services and national wealth are produced. This is particularly the case when the political economy becomes money-debt managed with the financial sector operating as if it is independent from the real sector. The financial assets are produced as commodities with small real investment where such production operates under supply-demand conditions in the markets in the financial sector. The conditions for the quantity of financial instruments supplied and demanded in the financial sector are different from the conditions of quantity of goods and services supplied and demanded in the real sector. There is, therefore, a built-in tendency to over-create financial instruments when the sector is not regulated or poorly regulated. The regulations in the financial sector must guide against increasing gambling in order to reduce the quantitative distance between the values of the real-sector production and financial-sector production for stability of the political economy. There are situations where restrictions on the financial instrument creation produce liquidity constraints and prevent the smooth functioning of the production-consumption process. The real-financial stability requires that the deviation between an index of financial assets in the financial sector measured in the index of real output and an index of real output in the real sector should be within a defined closeness for sustainability and stability of the political economy.

The stability may be specified as the epsilon-delta $(\varepsilon - \delta)$ stability condition. In other words, the financial bubble and the phantom wealth must be contained within a minimum of the maxima as the political economy moves over time. The objective of the governmental regulation in the financial sector is to constrain the value of financial growth within a reasonable distance from the value of real output.

It may be noted that the demand for financial instruments is not based on direct utility of consumption as in the case of the demand for real goods and services, but rather on indirect increased utility of consumption that financial capital gains and interest income provide with a possible risk of losses. This makes the price of a financial asset to depend on the interest income and riskiness and possible dividend, if any, that it is associated with, under conditions of fuzzy and stochastic uncertainties thus creating regulatory complexities to deal with them. The fuzzy uncertainty relates to vagueness of information and ambiguities of information signals that the financial markets generate while stochastic uncertainty relates to incomplete information that may be associated with market secrecy of all kinds where such information is available to all traders and buyers. The information inputs of decisions of the *financial bureaucratic firms* are different from those of *industrial bureaucratic firms* even though both sets of firms are driven by profit-making, enhanced by rent-seeking activities, thus creating additional complications for effective extensive and intensive regulations. The complexities of the regulatory regimes increase when there are inter-ownership relations among the financial and industrial bureaucratic firms that have power over the effects of the decisions of the decision-making core which creates and enforces regulations and laws.

4.1 The Financial Sector Conditions in the Schumpeterian Political Economy

Let us specify the basic conditions that will describe the financial sector, its rate of growth and expansion. In this respect, let $Q(t)$ be a measure of total real output at time t and the growth of output be $q(t) = \left(\frac{\dot{Q}}{Q}\right)$. Let the total financial assets be $F(t)$ at time t with its growth rate specified as $f(t) = \left(\frac{\dot{F}}{F}\right)$ all reduced into units of real output. For simplicity, let us divide the financial assets into three sub-aggregates that are made up of consumer indebtedness $E(t)$ with a growth rate $e(t) = \left(\frac{\dot{E}}{E}\right)$, housing indebtedness, $H(t)$ with a growth rate $h(t) = \left(\frac{\dot{H}}{H}\right)$, and corporate and government bonds $B(t)$ with its growth rate $b(t) = \left(\frac{\dot{B}}{B}\right)$. The aggregate growth rate of financial assets is an aggregative function of all its instruments. Let the initial conditions in the financial sector be specified as $F(t_0) = F_0$, $H(t_0) = H_0$, $E(t_0) = E_0$, $B(t_0) = B_0$. Furthermore, let the initial conditions of the endowment be constant with growth rates at those initial conditions being zero for all the inputs. We may set $E(0) = H(0) = B(0) = 1$. The three financial instruments may be seen in terms of quality in terms of risk, return, interest bearing and maturity period. Thus we may speak of the sets of bonds, consumer indebtedness and housing indebtedness, all of which may be specified as: $\mathbb{B} = \{B_i \mid i \in \mathbb{I}_B\}$, $\mathbb{E} = \{E_i \mid i \in \mathbb{I}_E\}$, and $\mathbb{H} = \{H_i \mid i \in \mathbb{I}_H\}$ respectively. We shall, however, work with the aggregate values of $\{B, E, \text{and } H\}$. Note that $\mathbb{B} = \{B_i \mid i \in \mathbb{I}_B\}$ contains business and government indebtedness.

The input-output relation in the financial sector is such that there is a total financial output $F(t)$ and a set of inputs $\{B(t), E(t), \text{and } H(t)\}$. The functional input-output relation may generally be described as:

$$F(t) = F(E(t), B(t), H(t)). \tag{4.1}$$

The function $F(\cdot)$ is a financial production function and it is taken to be a continuous function of its inputs with positive marginal contributions to the expansion of the financial output such that $F_E > 0$, $F_H > 0$ and $F_B > 0$. The inputs are continuous functions of time. For the purpose of the illustrative example, let the function be represented as a homogeneous function of the form:

$$F(t) = F(E(t), B(t), H(t)) = DE^{\alpha_1} H^{\alpha_2} B^{\alpha_3}, \ \alpha_1, \alpha_2, \alpha_3 \in [0,1), \sum_{i=1}^{3} \alpha_i. \tag{4.2}$$

The equation (4.2) may be complicated since, in general, the inputs of $\{B(t), E(t), \text{ and } H(t)\}$ are functions of other socioeconomic variables. In these discussions, the inputs are simply taken to be functions of time. The value $D(t)$ measures the social technology of the institutions of debt creation. Thus:

$$f(t) = \frac{\dot{F}}{F} = \frac{\dot{D}}{D} + \alpha_1 \frac{\dot{E}}{E} + \alpha_2 \frac{\dot{H}}{H} + \alpha_3 \frac{\dot{B}}{B}, \tag{4.3}$$

$$f(t) = \frac{\dot{F}}{F} = d + \alpha_1 e + \alpha_2 h + \alpha_3 b, \tag{4.4}$$

where $f(t) = \dfrac{\dot{F}}{F}$, $d = \dfrac{\dot{D}}{D}$, $e = \dfrac{\dot{E}}{E}$, $h = \dfrac{\dot{H}}{H}$, $b = \dfrac{\dot{B}}{B}$.   Equation   (4.4) describes the growth of the quantity of debt in the financial market given the behavior of the money market. It is a weighted average of the growth rates of the three markets augmented by the social technology of financial institutions that provide the structure of engineering of financial assets. It is this social technology of the institutions of debt creation that is under different regulatory regimes.

The structure of the money market is such that the government supplies money but holds no money. All the money stock in the system is held by the households, firms and lending institutions where $M_G = M_H + M_F + M_B + M_L$ at any time point where the M's are as specified in Table 3.1. It is useful for us in this analysis to distinguish between cash-liquidity in hand and non-cash-liquidity where liquidity is defined in the traditional sense. The stock of money is the basis of debt creation in terms of financial instruments that may be divided into liquid and non-liquid assets, the creations of which are under the control of the government through the legal structure. The total stock of money is always in the hands of the economic participants as the original holders or secondary holders in short-term transfers. The term $D(t)$ is an index of social technology that is composed of the institutional framework of financial markets, lending-borrowing conditions and conditions of regulations that link the financial sector to the real sector and allow an efficient management of the institutions of production and consumption with transfers of claims over wealth.

We may now investigate the rates of growth of the size of the value of financial output $f(t)$ and compare it with the rate of growth of the value of real output in the real sector. The speed with which the financial output grows may be written as:

$$\dot{f}(t) = \dot{d} + \alpha_1 \dot{e} + \alpha_2 \dot{h} + \alpha_3 \dot{b} \gtreqless 0. \tag{4.5}$$

Thus, we are looking for $\dot{f}(t)$ as specifying the nature of acceleration of the growth of debt in the political economy. The terms $(\dot{e}, \dot{h} \text{ and } \dot{b})$ specify the acceleration of the financial components in the debt market. The term $\dot{d}$ specifies the net speed of changes in the institutional constraint in terms of technology of the financial output, intensive and extensive regulations. The behavior of the system is such that the condition $\dot{f}(t) > 0$ implies that that technology of the financial output outweighs the negative effect of intensive and extensive regulations where $(\dot{e} + \dot{h} + \dot{b}) > 0$. On the other hand the condition $\dot{f}(t) < 0$ implies that the reversal process holds where the negative effect of regulation outweighs the technology of engineering the financial instrument. The condition $\dot{f}(t) = 0$, implies there is zero acceleration. As such, there is a balance between regulation and technology of the financial output with $(\dot{e} + \dot{h} + \dot{b}) = 0$ in which case the acceleration rates are compensatory in transfers where $\alpha_3 \dot{b} = -(\alpha_1 \dot{e} + \alpha_2 \dot{h})$.

The direction and the speed of $f(t)$ are constrained by politico-legal conditions of the political economy, and the shift of the grand ideological structure in the political economy regarding the role of the government and the behavior of the production entities in the financial sector. Under stable conditions of the ideological structure and the distribution of production entities, we shall assume a net balance of institutional acceleration so that $\dot{d} = 0$ and hence $d = \left(\dot{D}/_D\right)$ is constant. This will correspond to a situation where the financial system is traversing in the neighborhood of the steady state. In this respect, the financial components $(e, h \text{ and } b)$ are growing at a steady state rate with $(\dot{e} = \dot{h} = \dot{b} = g)$ so that given the initial conditions of the rates of growth of $(E, H \text{ and } B)$ as $(e_0, h_0 \text{ and } b_0)$, we have $(e = e_0 e^{gt}, h = h_0 e^{gt} \text{ and } b = b_0 e^{gt})$ so that given $(e, h \text{ and } b)$, $\dot{f} = (\alpha_1 + \alpha_2 + \alpha_3) g$ and hence:

$$f(t) = (\alpha_1 e_0 + \alpha_2 h_0 + \alpha_3 b_0) e^{gt} = f_0 e^{gt} \qquad (4.6)$$

where $f_0(0) = (\alpha_1 e_0 + \alpha_2 h_0 + \alpha_3 b_0)$ at $\dot{d} = 0$. We have assumed the accelerations to be equal for all the components of the debt output. It may be observed that equal accelerations do not mean equal growth rates. The structure of the debt sector and its dynamics with implicit income-risk profiles may be summarized as:

$$F(t) = F(E, H, B), \ F_E > 0, \ F_H > 0 \ F_B > 0 \qquad (4.7)$$

$$F(t) = F(E, H, B), \ DE^{\alpha_1} H^{\alpha_2} B^{\alpha_3} \qquad (4.8)$$

$$f(t) = \frac{\dot{F}}{F} = \frac{\dot{D}}{D} + \alpha_1 \frac{\dot{E}}{E} + \alpha_2 \frac{\dot{H}}{H} + \alpha_3 \frac{\dot{B}}{B} \qquad (4.9a)$$

$$f(t) = d + \alpha_1 e + \alpha_2 h + \alpha_3 b \qquad (4.9b)$$

$$f(t) = \left(\alpha_1 e_0 + \alpha_2 h_0 + \alpha_3 b_0 \right) e^{gt} = f_0 e^{gt}. \qquad (4.10)$$

The system of equations (4.7 - 4.10) defines the general dynamic debt effects in the political economy where $\alpha_i \in [0,1]$ are debt issuance distribution parameters. To complete the total financial effects in the political economy, we must add the financial coupons held in the money market to obtain the total financial claims over real commodities in the real sector.

4.1.1 The Relationship between the Money and Debt Markets

To examine the effects of total financial-sector claims on the real sector outputs, we need to relate the debt market to the money market. Let us return to the T-account of the consolidated national wealth-and-financial claims in the closed economy and examine how the two markets in the financial sector are inter-supportive in their relationship. Table 3.1 is not a balance sheet but a wealth-claims account as a technique in understanding macroeconomic relations in the political economy regarding real income-wealth distribution, financial claims and rent-seeking activities. The total financial claims $\mathbb{F}$ in a simple form, is the monetary base $\mathbb{B}$ plus total debt $\mathbb{D}$ in the closed economy. The monetary base as financial claims comes from the money market while the debt instruments as financial claims come from the debt market. The maximum value of debt or loan that can be issued in the system depends on the regulatory regime, reserved-requirement ratio and the nature of the bureaucratic financial innovation process that must be seen in terms of rent-seeking and profit enhancing. If $\vartheta \in [0,1]$ is the reserved requirement ratio, then the total potential financial claims through debt creation is $\mathbb{D} = k\mathbb{B}$ and $k = (1/\vartheta)$, where k is a money multiplier depending on the regulatory regime and the total financial claims is

$$\mathbb{F} = \mathbb{B} + \mathbb{D} = \mathbb{B} + k\mathbb{B} = (1+k)\mathbb{B}. \qquad (4.11)$$

$$F(t) = F(E, H, B), \ DE^{\alpha_1} H^{\alpha_2} B^{\alpha_3} \leq k\mathbb{B}(t) \qquad (4.12)$$

From equation (4.11) $\mathbf{f}(t) = \dot{\mathbf{F}} = \dot{\mathbf{B}} + \dot{\mathbf{D}} = \dot{\mathbf{B}} + k\dot{\mathbf{B}} = (1+k)\dot{\mathbf{B}}$ and the acceleration of the growth of the value in the financial sector composed of the money and debt markets is $\dot{\mathbf{f}}(t) == (1+k)\ddot{\mathbf{B}} >$ $f(t) = (\alpha_1 e_0 + \alpha_2 h_0 + \alpha_3 b_0)e^{gt} = f_0 e^{gt}$.

The acceleration of the growth of the volume of the debt market is then dependent on the acceleration of the growth of the monetary base and the money multiplier which is institutionally determined to ensure financial stability and sustainable financial risk to the political economy. The control of the money multiplier is complemented by institutional controls in terms of the criteria to belong to the debt issuance system. Given the multiplier and the institutions, the stability in the financial system depends on the intensive and extensive regulations that restrict financial activities and profits in the financial sector. Given the volume of the financial assets in the debt market, the financial stability may be compromised by active trading and selling of debt where capital gains may be abstracted by the manipulations of the present, forward and future prices of debt instruments under a fuzzy-stochastic environment. These financial papers have no value except that they hold and guarantee claims over the real wealth. To examine the stability of the political economy, the total financial claims must be related to the total real commodity claims in the real sector. As such let us visit the real sector production system.

4.2   The Real-Sector Conditions in the Schumpeterian Political Economy

Our interest in the theory of the political economy as it relates to the rent-seeking phenomena and income-wealth distribution is on the value relation between the real sector and the financial sector. The role of the financial sector has already been examined. Let us now turn our attention to the role of the real sector in the political economy and specify the conditions that will describe it, its output, the rate of growth of output and the acceleration of the growth rate. Let $Q(t)$ be the measure of real total output in a closed economy at time (t) and corresponding to it a growth path of $q = \left(\dfrac{\dot{Q}}{Q}\right)$ with acceleration $\dot{q}$. The real sector dynamics are generated by a labor-capital process that, for illustrative conditions, may be specified simply by linearly homogeneous types of the Cobb-Douglas production system with lineal homogeneous functions, where K is a flow of capital service input, L is a flow of labor service input and A is a technological variable representing the social and physical processes of knowledge production. The real sector dynamics are composed of eqns. (4.13 – 4.15) with initial conditions $Q(t_0) = Q_0$, $L(t_0) = L_0$, $K(t_0) = K_0$ and with a given social technology of production that varies with institutional learning. The use of linearly homogeneous functions allows us to always close an income-wealth-distribution closure of the political economy.

$$Q(t) = F(K, L) = AK^{\beta_1}L^{\beta_2} , \quad (0 < \beta_1, \beta_2 <, \; \beta_1 + \beta_2 = 1) \tag{4.13}$$

The rate of growth at any time may be specified as

$$q(t) = \frac{\dot{Q}}{Q} = \frac{\dot{A}}{A} + \beta_1 \frac{\dot{K}}{K} + \beta_2 \frac{\dot{L}}{L} = a + \beta_1 k + \beta_2 l \gtrless 0 \tag{4.14}$$

From equation (4.15) we may specify the acceleration rate as:

$$\dot{q}(t) = \left(\dot{a} + \beta_1 \dot{k} + \beta_2 \dot{l} \right) \gtrless 0 \tag{4.15}$$

where $a = \left(\dfrac{\dot{A}}{A} \right)$, $\; k = \left(\dfrac{\dot{K}}{K} \right)$, and $\; l = \left(\dfrac{\dot{L}}{L} \right)$

In the neoclassical setting β_1 and β_2 are technologically determined in the sense that the acknowledgement of contributions of the rates of growth of factors is technologically determined by the physical technological system. In the Neo-Keynesian setting, β_1 and β_2 are institutionally determined in the sense that the acknowledgment of the contributions of the rates of growth of factors to the overall growth is placed on the institutional configuration that holds the social technological system. The differential acknowledgements by these two schools of economic thought in the nature of the political economy and income-wealth distribution are extremely important to the economic policy debates and the whole of the incentive structure that affects the social income-wealth distribution in the real sector [R2.10] [R2.17][R2.22][R2.36] [R2.47] [R2.92] [R2.120] [R2.127]. The structure of incentives and the pattern of remuneration in the political economy may thus be taken as either institutionally or technologically determined in a mutual exclusivity or non-mutual exclusivity, where mutual exclusivity involves the extremes and the non-mutual exclusivity involves the combination of physical and social technologies. The assumed nature of income determination is analytically important in examining the dynamics of consumption-savings balances and the resulting real wealth available in the social system for the understanding of the dynamics of borrowing-lending activities in the financial sector.

For a stable process and idea that technological and institutional changes take time and cannot be speed up, we take (a) to be constant and hence, $(\dot{a}=0)$. Thus:

$$\dot{q}(t)=\left(\beta_1\dot{k}+\beta_2\dot{i}\right)\gtreqless 0 \tag{4.16}$$

In other words, given an institutional socio-technological stable growth, the acceleration of real output growth depends on the accelerations of the growth rates of flows of the capital and labor services in the real sector and the activities in the factor market. The acceleration may be positive $\dot{q}(t)>0$, zero $\dot{q}(t)=0$ or negative $\dot{q}(t)<0$ which must be related to both real output and the financial system's dynamics. The real out-put dynamics must be related to factor-usage dynamics in terms of the rates of employment increases and capital-capacity utilization. The financial dynamics must be related to the rates of increases in the money supply and financial assets. The condition $\dot{q}(t)>0$ implies that the net acceleration of factor services of capital and labor is positive increasing rate of employment, and integration of capital service usage. Similarly, the condition $\dot{q}(t)<0$ implies that the net acceleration of factor services of capital and labor is negative (reduction in the rate of net factor usage in production). The condition $\dot{q}(t)=0$ implies that the net acceleration of factor services of capital and labor is zero. This may be accompanied by compensation variation in addition to others. Thus there are three cases of acceleration to be examined.

Case I: $\dot{q}(t)>0$, $\beta_1<\beta_2$ and $(\beta_1+\beta_2)=1$, $(\beta_1,\beta_2)\in[0,1]$

   a) If $\dot{k}>0$, $\dot{i}>0$ then $\left(\beta_1\dot{k}+\beta_2\dot{i}\right)>0$ (4.17)

(Increasing rate of both employment and capital-service usage)

   b) $\dot{q}(t)>0$, $\dot{k}<0$, and $\dot{i}>0$ then $\left(\beta_1\dot{k}+\beta_2\dot{i}\right)>0$ $\dot{i}>\frac{\beta_1}{\beta_2}\dot{k}$ (4.18)

(Expansion in labor usage through labor substitution and increased employment

   c) $\dot{q}(t)>0$, $\dot{k}>0$, and $\dot{i}<0$ then $\left(\beta_1\dot{k}-\beta_2\dot{i}\right)>0$ $\dot{k}>\frac{\beta_2}{\beta_1}\dot{i}$.   (4.19)

(Expansion in capital usage through capital substitution and decreasing labor employment)

Case II: $\dot{q}(t) < 0$, $\beta_1 < \beta_2$ and $(\beta_1 + \beta_2) = 1$, $(\beta_1, \beta_2) \in [0,1]$

$$\text{a) If } \dot{k} < 0,\ \dot{i} < 0 \ \Rightarrow\ \left(\beta_1 \dot{k} + \beta_2 \dot{i}\right) < 0. \tag{4.20}$$

(Decreasing rates of both the rates of employment and capital-capacity usage)

$$\text{b) } \dot{q}(t) < 0,\ \dot{k} > 0,\ \text{and } \dot{i} < 0 \text{ then } \left(\beta_1 \dot{k} - \beta_2 \dot{i}\right) < 0 \ \ \dot{k} < \tfrac{\beta_2}{\beta_1}\dot{i}. \tag{4.21}$$

(Capital substitution and decreasing rate of labor employment)

$$\text{c) } \dot{q}(t) < 0,\ \dot{k} < 0,\ \text{and } \dot{i} > 0 \ \Rightarrow\ \left(\beta_1 \dot{k} + \beta_2 \dot{i}\right) < 0 \ \ \dot{i} > \tfrac{\beta_1}{\beta_2}\dot{k}. \tag{4.22}$$

(Employment expansion through labor substitution and decreasing rate of capital usage)

Case III $\dot{q}(t) = 0$, $\beta_1 < \beta_2$ and $(\beta_1 + \beta_2) = 1$, $(\beta_1, \beta_2) \in [0,1]$

$$\text{a) If } \left(\beta_1 \dot{k} + \beta_2 \dot{i}\right) = 0 \ \Rightarrow\ \dot{k} = \dot{i} = 0, \text{ since } (\beta_1, \beta_2) \in [0,1]. \tag{4.23}$$

(The rates of employment and capital-capacity utilization stay the same)

$$\text{b) } \beta_1 \dot{k} = -\beta_2 \dot{i} > 0 \ \ \dot{k} > -\tfrac{\beta_2}{\beta_1}\dot{i} \Rightarrow \text{ employment reduction}. \tag{4.24}$$

$$\text{c) } \beta_2 \dot{i} = -\beta_1 \dot{k} \ \ \dot{k} > -\tfrac{\beta_2}{\beta_1}\dot{i} \Rightarrow \text{ capital reduction}. \tag{4.25}$$

Equations (4.16 – 4.25) constitute the system's dynamics in the real sector that must be coordinated with the financial-sector dynamics and the consumption-savings processes in the political economy. In Chapter three of this monograph, initial discussions were provided regarding national wealth distribution. We now turn our attention to the real output distribution in the factor market and then relate it to the financial sector activities and financial asset holdings.

4.3 Conditions of Output Distribution as Institutional Dynamics in the Schumpeterian Political Economy

The dynamic behavior of the political economy requires three sets of dynamic equations of the real sector, the financial sector in the economic structure and institutional support from the political and legal structures. Given the institutional framework, the set of equations of motion of the financial and real sector dynamics must be supported by a set of equations of the income distribution and consumption-savings behavior of the society in the closed political economy. The income-distribution and consumption-savings dynamics are representations of the institutional dynamics that contribute to the understanding of the borrowing-lending process and the regulatory dynamics. The dynamics of income distribution and the consumption-savings process take place in the real sector through the money market, while the lending-borrowing process also takes place in the real sector but through the debt market which affects the claim-transfer process for consumption and wealth ownerships.

The set of equations of motion of the institutional dynamics may be developed from the notion that the value of total output $Q(t)$, at anytime, is institutionally distributed between wages $W(t)$ and profits $\Pi(t)$ so that:

$$Q(t) = \Pi(t) + W(t). \tag{4.26}$$

In this formulation, $\Pi(t) = rK(t)$ and $W(t) = wL(t)$, where $r = \left(\frac{R}{P}\right)$ and $w = \left(\frac{W}{P}\right)$ as the real unit costs of capital and labor respectively, where the measure of aggregate price P is valued at one. R and W are nominal unit values of capital and labor respectively. Similarly, the system is such that total output is divided into consumption $C(t)$ and savings $S(t)$ at any time. Thus:

$$Q(t) = C(t) + S(t). \tag{4.27}$$

The total savings at any time t goes to increase real national wealth $\mathbb{W}(t)$ which is an accumulation savings from any initialized period. Thus:

$$\left. \begin{aligned}
\mathbb{W}(t) &= \int_{-\infty}^{t} S(t)\, dt \\
\dot{\mathbb{W}}(t) &= S(t) \\
S(t) &= s_\Pi \Pi(t) + s_w \mathbb{W}(t), \quad (s_\Pi, s_w) \in [0,1] \\
S(t) &= s_\Pi rK(t) + s_w wL(t) \\
s(t) &= \gamma_1 s_\Pi + \gamma_2 s_w, \quad (\gamma_1, \gamma_2) \in [0,1], \; (\gamma_1 + \gamma_2) = 1
\end{aligned} \right\} . \tag{4.28}$$

The condition $\dot{\mathbb{W}} < 0$ implies that there is a reduction in the existing real national wealth while the condition $\dot{\mathbb{W}} > 0$ implies an increase in the existing real national wealth. The value $S(t)$ is an addition to real loanable funds which then depends on income distribution and savings habits of the respective income classes. The total increase in total savings to the national wealth is:

$$S(t) = sQ(t) = \dot{\mathbb{W}}(t). \tag{4.29}$$

In this political economy, total real savings does two things. It is equal to the increase in the national wealth which is an increase in real aggregate inventory that can be used for investment through the real lending-borrowing process in the real sector.

We may now examine the acceleration process in terms of the rates of growth in the real sector for the growth of real wealth. The acceleration rate of growth of the real funding process in terms of demands for consumption and investment goods and services for any culture of real income distribution may be specified as:

$$
\left.\begin{aligned}
\dot{S}(t) &= s_\Pi \dot{\Pi}(t) + s_w \dot{W}(t) \\
s\dot{Q} &= s_\Pi r K\left(\frac{\dot{r}}{r} + \frac{\dot{K}}{K}\right) + s_w w L\left(\frac{\dot{w}}{w} + \frac{\dot{L}}{L}\right) = s_\Pi r K\left(\frac{\dot{r}}{r} + k\right) + s_w w L\left(\frac{\dot{w}}{w} + l\right) \\
sq &= s\frac{\dot{Q}}{Q} = s_\Pi r \frac{K}{Q}\left(\frac{\dot{r}}{r} + k\right) + s_w w \frac{L}{Q}\left(\frac{\dot{w}}{w} + l\right)
\end{aligned}\right\} \qquad (4.30)
$$

In equation (4.30) $k = \left(\dfrac{\dot{K}}{K}\right)$ and $l = \left(\dfrac{\dot{L}}{L}\right)$. Let $v = \dfrac{Q}{K}$ be a stable value

of average capital productivity and $\lambda = \dfrac{Q}{L}$ be also a stable measure of average labor productivity over a given culture of production and technical knowhow. Then we can write:

$$
sq = s\frac{\dot{Q}}{Q} = \frac{s_\Pi r}{v}\left(\frac{\dot{r}}{r} + k\right) + \frac{s_w w}{\lambda}\left(\frac{\dot{w}}{w} + l\right) \qquad (4.31)
$$

To examine the interactive nature of the real and financial sectors, the real sector must be viewed as providing collaterals for the legitimacy and people's willingness to accept money and credit through the debt process. The relationship must be seen in terms of a moving equilibrium and a stable moving disequilibrium. The equilibria and stability of the relationship between the real and financial sectors are governed by the two equations of motion that describe the growth of financial holdings in the financial sector, eqn. (4.32) and the growth of real wealth in the real sector (4.33) which are summaries from above. These growth equations are supported by social distribution of real income additions to the national wealth as specified in eqns. (4.34) and (4.35)

$$
f(t) = \alpha_0 + \alpha_1 e + \alpha_2 h + \alpha_3 b \qquad (4.32)
$$

$$
q(t) = a + \beta_1 k + \beta_2 l \qquad (4.33)
$$

$$sq = \frac{S_\Pi r}{v}\left(\frac{\dot{r}}{r}+k\right) + \frac{S_w w}{\lambda}\left(\frac{\dot{w}}{w}+l\right) \tag{4.34}$$

$$q = \frac{S_\Pi r}{vs}\left(\frac{\dot{r}}{r}+k\right) + \frac{S_w w}{\lambda s}\left(\frac{\dot{w}}{w}+l\right) \tag{4.35}$$

Let the average productivity of capital and labor be constants where

$S_{\Pi_0} = \left(\frac{(s_\Pi r)}{(sv)}\right)$ defines initial savings conditions of profit from capital and

$S_{W_0} = \left(\frac{(s_w w)}{(s\lambda)}\right)$ defines the initial saving conditions from the growth of wage income. Given $\left(\frac{\dot{r}}{r}\right) = \rho$ and $\left(\frac{\dot{w}}{w}\right) = \omega$, the growth of output is related to the growth of factor remuneration and the growth of factor intensities in the political economy as specified in equation (4.36).

$$q = S_{\Pi_0}\left(\rho+k\right) + S_{W_0}\left(\omega+l\right) \tag{4.36}$$

The rate of acceleration is given as:

$$\dot{q} = S_{\Pi_0}\left(\dot{\rho}+\dot{k}\right) + S_{W_0}\left(\dot{\omega}+\dot{l}\right) \tag{4.37}$$

Equations (4.33 -4.37) constitute the dynamics of real sector $\mathcal{R}$. We shall refer to $S_{\Pi_0}\left(\dot{\rho}+\dot{k}\right)$ as the *real capital effect* on the speed of the rate of real wealth growth, and $S_{W_0}\left(\dot{\omega}+\dot{l}\right)$ as the real *labor effect* on the speed of the rate of real wealth growth given their initial conditions, and since the real output is growing at a decreasing rate, we postulate a fuzzy approximation of the combined effect as an exponentially decreasing function with $g = \left(\dot{\rho}+\dot{k}\right) = \left(\dot{\omega}+\dot{l}\right)$ and $S_0 = \left(S_{\Pi_0} + S_{W_0}\right)$ in the proximity of the steady state

where $\qquad \dot{q} = S_{\Pi_0}\left(\dot{\rho}+\dot{k}\right) + S_{W_0}\left(\dot{\omega}+\dot{l}\right) \le 0 \Rightarrow \left(\frac{S_{\Pi_0}}{S_{W_0}}\right) \le \left(\frac{\dot{w}+\dot{l}}{\dot{\rho}+\dot{k}}\right) \qquad$ where S_{Π_0} and S_{W_0} are distributive weights.

$$\dot{\sigma} = sq = \left(S_{\Pi_0} + S_{W_0}\right)e^{-gt} = S_0 e^{-gt} \tag{4.38}$$

$$f(t) = \left(\alpha_0 + \alpha_1\dot{e} + \alpha_2\dot{h} + \alpha_3\dot{b}\right) = F_0 e^{gt} \tag{4.39}$$

$$\dot{q}(t) = \left(\dot{a} + \beta_1\dot{k} + \beta_2\dot{l}\right) = \overline{S}_0 e^{-gt} \tag{4.40}$$

The values f, q, sq tell us the rates of growth of the aggregate financial instrument, real output and real savings respectively. The values $\dot{f}, \dot{q}, \dot{\sigma} = s\dot{q}$ provide us with information on how fast (acceleration) the rates of growth of the aggregate financial instrument, real output and real savings are moving respectively. Thus, we have a derived secondary dynamic system for the relationship between the financial sector and the real sector represented as:

$$\dot{f}(t) = F_0 e^{gt} . \tag{4.41}$$

$$\dot{\sigma} = S_0 e^{-gt} . \tag{4.42}$$

We can now introduce in the analysis, the concept and measurement of the size of the *financial bubble* $\mathscr{B}$ and its time rate of growth $\dot{\mathscr{B}}$

$$\mathscr{B} = f(t) - q(t) = \left(\alpha_0 + \alpha_1 e + \alpha_2 h + \alpha_3 b \right) - \left(\alpha + \beta_1 k + \beta_2 l \right) . \tag{4.43}$$

$$\dot{\mathscr{B}} = \dot{f}(t) - \dot{q}(t) = \left(\alpha_0 + \alpha_1 \dot{e} + \alpha_2 \dot{h} + \alpha_3 \dot{b} \right) - \left(\alpha + \beta_1 \dot{k} + \beta_2 \dot{i} \right) . \tag{4.44}$$

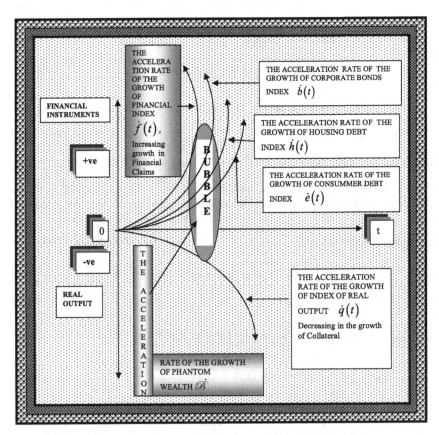

Fig. 4.2 The Geometry of Phantom Wealth Creation and Chaotic Financial System

Equation (4.43) defines the dynamic path (growth) of a disparity between the real and financial sectors while equation (4.44) defines the rate (acceleration) at which this disparity is changing in the political economy. The disparity defines a financial bubble and the rate at which the bubble is increasing is the acceleration of the disparity. A geometric structure of these equations is provided in Figure (4.2). The following definitions are useful in understanding the dynamics of the real-financial sector duality and the problems of financial instability.

Definition 4.1

The economic structure $\mathbb{E}$ of a political economy $\mathscr{P}$ is a system defined by a Cartesian product of the form $\mathbb{E} = \left(\mathscr{F} \otimes \mathscr{R} \right)$ for any given population size and government where $\mathscr{F}$ and $\mathscr{R}$ are sets of financial output and real output configurations given the political and legal structures $\mathbb{P}$ and $\mathbb{L}$ respectively.

The symbol $\mathscr{F}$ is a set of financial production attributes and $\mathscr{R}$ is a set of real production attributes. If I is an index set of elements in $\mathscr{F}$ with a generic element, $i \in \mathsf{I}$ and J is an index set of the elements in $\mathscr{R}$ with a generic element $j \in \mathsf{J}$, then we may specify the political economy as a set of relationships of the financial and real sector elements as:

$$\mathbb{E} = \left(\mathscr{F} \otimes \mathscr{R} \right) = \left\{ e = \left(f_i, r_j \right) \mid f_i \in \mathscr{F}, i \in \mathsf{I}, r_j \in \mathscr{R} \text{ and } j \in \mathsf{J} \right\} \quad (4.45)$$

Definition 4.2

The financial sector $\mathscr{F}$ is said to be completely collateralized if every element $f_i \in \mathscr{F} \; \exists \; r_j \in \mathscr{R} \; \forall i \in \mathsf{I}$ and $j \in \mathsf{J}$ that covers its value. It is said to be partially collateralized if for some $f_i \in \mathscr{F} \; \not\exists \; r_j \in \mathscr{R}, i \in \mathsf{I}$ and $j \in \mathsf{J}$ that covers its value. A financial bubble $\mathscr{B}$ exists in the political economy if the financial sector is partially collateralized and hence $\mathscr{B} > 0$. Every financial sector is a Cartesian product of money-market products, $\mathbb{M}$ and debt-market products, $\mathbb{D}$ of the form $\mathscr{F} = \left(\mathbb{D} \otimes \mathbb{M} \right)$. Similarly, every real sector $\mathscr{R}$ is a Cartesian product of the form $\mathscr{R} = \left(\mathbb{F} \otimes \mathbb{C} \right)$ where $\mathbb{F}$ is the factor market products, and $\mathbb{C}$ is commodity market products.

Note 4.1

The political economy $\mathscr{P}$ is defined as a Cartesian triplet of configurations of political $\mathbb{P}$, economic $\mathbb{E}$, and legal $\mathbb{L}$ structures with $\mathscr{P} = \left(\mathbb{P} \otimes \mathbb{E} \otimes \mathbb{L} \right)$. As defined, the political economy is also the social formation. We may also note that the notion of credit creation on the principle of proportionality of high powered money is the foundation of a financial bubble and the rise of a phantom wealth. As the two sectors stand, they seem to be separately managed without some organic tie to their manipulations even though the real sector collateralizes the financial sector without which the financial production has no use value. The value of the financial bubble, its growth and new loans at any time point is collateralized by future production and its growth. The implication is that the expectations of future production and the possible rate of growth will help to determine the sustainability of the bubble and its rate of growth as well as the systemic risk of the political economy.

Proposition 4.1

A political economy $\mathscr{P}$ is in complete equilibrium if $\mathscr{B} = 0$ and $\dot{\mathscr{B}} = 0$ and hence the financial-sector production is fully collateralized by the real-sector production. It is said to be in disequilibrium if the financial sector is partially collateralized. It is on an $(\varepsilon - \delta)$ stable path if $\dot{\mathscr{B}} \neq 0$, then given a positive δ, there exists an (ε) such that when $\left(\left|\mathscr{F} - \mathscr{R}\right|\right) \leq \delta$ or $\left|\mathscr{B}\right| \geq 0$ then $\left|\dot{\mathscr{B}}\right| \leq \varepsilon$.

The complete equilibrium between the real and financial sectors will be called real-financial sector equilibrium. Deviations from this equilibrium where the financial sector is partially collateralized will be called real-financial sector disequilibrium that generates a financial *systemic risk* for the political economy by the nature of the principle of fractional reserved banking in debt creation for generating financial claims over elements in the real sector. The bubble, the phantom wealth and the financial systemic risk are built into the economic structure and managed from the political and legal structures through rules and regulations. Definition 4.1 and proposition 4.1 simply suggest to us that the stability of the political economy, composed of interactions of the real and financial sectors, can be managed through regulations that impose control mechanisms to deal with the ε and δ in order to restrict the distance between the aggregate index values of the real and financial sectors within socially acceptable ranges and hence the size of the bubble is optimally controlled with a systemic risk that is organizationally allowed. The system is such that, given a measure of δ from the system's behavior, the government must institute regulatory measures to create ε-value to control the bubble within limits, and thus manage the sustainable *systemic risk* of the political economy. The managed sustainable bubble is then collateralized by future real production, the value of which is stochastically fuzzy which requires judicious policy through the use of subjective and objective information. It is here, that the relational structure of information and control decisions becomes critically important in the managing of the dynamics of the political economy. It is also here that the growth of the financial bubble becomes the growth of wealth transfers in the political economy in the private sector. Such wealth transfers can be hastened and amplified through rent-seeking activities by the enterprises in the financial sector.

4.4 The Phantom Wealth, Financial Bubbles, Rent-Seeking Activities and Financial Instabilities in the Schumpeterian Political Economy

The real-financial sector equilibrium will require that, give the initial equilibrium, the time value of the growth of real output should be equal to the growth of real value of financial output, where $q(t) = f(t)$ given the initial conditions and the unit of valuation and hence:

$$q = \frac{s_\Pi r}{vs}\left(\frac{\dot{r}}{r}+k\right) + \frac{s_w w}{\lambda s}\left(\frac{\dot{w}}{w}+l\right) = \alpha_0 + \alpha_1 e + \alpha_2 h + \alpha_3 b \qquad (4.46)$$

The conditions for stable equilibrium require that consumption lending-borrowing should balance with labor income; and similarly, investment lending-borrowing should balance with capital income under conditions of no taxation which can be relaxed. These require that from eqts. (4.35) and (4.43):

$$\frac{s_\Pi r}{vs}\left(\frac{\dot{r}}{r}+k\right) = \alpha_0 + \alpha_3 b = a + \beta_1 k . \qquad (4.47)$$

And

$$\frac{s_w w}{\lambda s}\left(\frac{\dot{w}}{w}+l\right) = \alpha_1 e + \alpha_2 h = \beta_2 l . \qquad (4.48)$$

Thus we can solve for the rates of capital-service and labor-service flows in terms of debt market and factor market variables to obtain capital labor conditions for equilibrium between real and financial sectors. Hence we have:

(I) Capital growth conditions and indebtedness for real-financial sector equilibrium:

$$k = \frac{\alpha_0 + \alpha_3 b - a}{b} \quad \text{debt market condition}$$

or

$$k = \frac{\left(s_\Pi \dot{r} - avs\right)}{\left(\beta_1 vs - s_\Pi r\right)} \quad \text{factor market condition} \qquad \left.\begin{array}{c} \\ \\ \\ \\ \\ \\ \\ \\ \\ \end{array}\right\} , \text{ for capital growth (4.49)}$$

and the growth of capital indebtedness is

$$b = \left[\frac{\left(\alpha_0 - \alpha\right)\left(\beta_1 vs - s_\Pi r\right)}{s_\Pi\left(\dot{r} - \alpha_3 r\right) - vs\left(\alpha + \alpha_3 \beta_1\right)}\right] , \text{ growth of bond}$$

The growth of capital services depends on debt conditions of private and public bonds in the financial sector while it depends on the rate of capital earnings and average productivity in the real sector given the savings culture from capital. By combining the two we can derive an expression for the growth of capital and government indebtedness in terms of the two conditions for equilibrium and stability as in the third expression in equation (4.49).

(II) Labor growth conditions for real-financial sector equilibrium:

$$l = \frac{\alpha_1 e + \alpha_2 h}{\beta_2} \quad \text{debt market condition}$$

or

$$l = \frac{s_w \dot{w}}{(\beta_2 \lambda s - s_w w)} \quad \text{factor market condition}$$

$$\left.\begin{array}{c} \\ \\ \end{array}\right\} \text{for labor growth (4.50)}$$

and the growth of labor indebtedness is

$$h = \left[\frac{\beta_2 s_w \dot{w}}{(\beta_2 \lambda s - s_w w)(\alpha_1 \sigma + \alpha_2)}\right], \text{ growth of housing ondebtedness}$$

The labor services growth depends on debt conditions of housing and consumer indebtedness as an approximation of labor indebtedness in the financial sector while it depends on the rate of labor earnings in the real sector and average productivity. By combining the two conditions we can derive an expression for the growth of labor indebtedness as the third expression in equation (4.50). The value $e = \sigma h$ is the growth rate of consumer indebtedness which is taken to be a stable proportion of increases in the housing indebtedness. In this formulation, aggregate consumer indebtedness will be found to be positively related to aggregate housing indebtedness when the data is examined.

4.4.1 Rent-Seeking, Real-Financial-Sector Disequilibrium and the Trading of Phantom Wealth in the Schumpeterian Political Economy

The economy is continually at disequilibrium between the real and financial sectors. The real-financial sector disequilibrium may come from either the money market in terms of the *primary money supply* or the debt market in terms of the *secondary money supply* through the growth of debt instruments or both. The major elements of disequilibrium come from the debt market. We shall assume that the primary money supply in the money market always satisfies the output growth in the real sector at each point of time. It is useful to keep in mind how the money supply enters into the financial sector. The lending-borrowing activities in the debt market are through fractional banking where such fractional condition is mandated from the legal structure through the political structure. Thus from eqn.(4.11), we have $\mathbb{F} = \mathbb{B} + \mathbb{D} = \mathbb{B} + k\mathbb{B} = (1+k)\mathbb{B}$ where $\mathbb{F}$ is the total debt, $\mathbb{B}$ is the monetary base as the primary money from the money market and $\mathbb{D}$ is the secondary money from the debt market, where $k = (\frac{1}{\vartheta}) > 1$ and k is the money multiplier for debt creation, and ϑ is the reserved requirement ratio as mandated by law. The fractional banking creates initial real-financial-sector disequilibrium with a defined bubble.

Since $k = \left(\frac{1}{\vartheta} \right)$ is fixed by law and $\mathbb{B}$ is under the control of the governmental debt creation in terms of the central bank notes, the business in the debt market recreates itself through different levels of ingenuity. The maximum supply of debt is $k\mathbb{B}$ for any given monetary base, $\mathbb{B}$. The debt market transforms itself from simple lending-borrowing activities into debt-trading instruments through the methods of financial engineering for capital gains. In other words, the debt market becomes transformed into an effective gambling space where those who know the game or are willing to learn the game can participate for gain-loss outcomes. Each commoditized debt instrument is seen as a coupon that has a price. Each coupon has a differential quality specified in terms of *interest-rate earnings* (or rate of return), *terms of maturity* and a *degree of risk*. It is then ranked in grades and initially priced for buying and selling. Let us keep in mind that each coupon, as a debt instrument, is sought after because it is believed to have collateral in the real sector in terms of wealth support.

The value of the financial instruments is the value of primary money creation plus the value of secondary money creation in a given monetary unit. The value of the produced assets in the real sector is the sum of each commodity converted to the monetary unit by their corresponding prices. The real sector is $\mathscr{R} = (\mathbb{F} \otimes \mathbb{C})$ where we can specify the real sector as composing of wealth for the real-sector markets, where $i \in \mathbb{I}$ is an index set of wealth of the closed national economy in the form $\mathscr{R}(t) = \left\{ x_i(t) = F_i \left(f_i(t), c_i(t) \right) \mid f \in \mathbb{F}, c \in \mathbb{C} \text{ and } i \in \mathbb{I} \right\}$, and t is a time element. The value of increased wealth $\dot{\mathbb{W}}(t)$ at each time is $\dot{\mathbb{W}}(t) = \sum_{i \in \mathbb{I}} p_i x_i = \mathbb{P}Q$. The values $\mathbb{P}(t)$ and $Q(t)$ are information indexes on the aggregate price and aggregate output respectively, and p_i and x_i are individual prices and commodities respectively. In this respect the equilibrium between the real sector and the financial sector is such that if we start with an initial equilibrium $\mathbb{P}(t_0)Q(t_0) = \mathbb{B}(t_0)$, then the real-financial-sector disequilibrium is generated by fractional banking in the form:

$$\mathbb{P}(t_0)Q(t_0) < (1+k)\mathbb{B}(t_0) \Rightarrow \frac{\mathbb{P}(t_0)Q(t_0)}{(1-k)} < \mathbb{B}(t_0) < (1+k)\mathbb{B}(t_0). \quad (4.51)$$

To maintain the initial equilibrium at each point of time, the aggregate price must rise to equilibrate the two values. The level of the aggregate price at the equilibrating process is $\mathbb{P}(t) = \dfrac{(1+k)\mathbb{B}(t)}{Q(t)}$, where k is the velocity of debt-creation. In this respect, the total nominal value of commodities at any time (t) must be equal to the nominal value of purchasing power in the system where the purchasing power is determined by the monetary base plus the legally allowable debt creation. The inflation in the aggregate wealth may be specified as:

$$\dot{P}(t) = \frac{\mathbb{B}(t)}{Q(t)} \left[(1+k) \frac{\dot{\mathbb{B}}(t)}{\mathbb{B}(t)} - (1+k) \frac{\dot{Q}(t)}{Q(t)} \right] = \mathbf{m}(1+k) [\mathbf{b} - q] \qquad (4.52)$$

The values are such that $\mathbf{m} = \frac{\mathbb{B}(t)}{Q(t)}$ is money-output ratio, $\mathbf{b} = \frac{\dot{\mathbb{B}}(t)}{\mathbb{B}(t)}$ is the growth of the monetary base, and $q = \frac{\dot{Q}(t)}{Q(t)}$ is the output growth rate. Thus the price inflation in the political economy depends on money-output ratio, the growth of money and the output growth on the path of the real-financial sector equilibrium. The implication is that the existing prices of real assets and the future prices of real commodities must rise on the aggregate. The value relation between the real-sector output and financial-sector output specified as $P(t)Q(t) = (1+k)\mathbb{B}(t)$ may be related to the Classical quantity theory of money. The stability condition of the real-financial interactive process requires that the growth of the monetary base be set to be equal to the growth of real-sector output. In an open economy $Q(t)$ may be redefined as $\hat{Q}(t)$ to include net imports $(X(t) - M(t))$, where $\hat{Q}(t) = Q(t) + (M(t) - X(t))$, $X(t)$ is exports and $M(t)$ is imports in real commodities and hence $P(t) \left[Q(t) + (M(t) - X(t)) \right] = (1+k)\mathbb{B}(t) \cdot$ In this case, we have $P(t) = \dfrac{(1+k)\mathbb{B}(t)}{\left[Q(t) + (M(t) - X(t)) \right]}$ where $(M(t) - X(t)) > 0$ implies addition to inventories, and balanced with addition to domestic international indebtedness. The price stability in the increases in financial instruments may be accomplished to the extent to which $(M(t) - X(t)) \gg 0 \cdot$ Generally, the increase $(1+k)\mathbb{B}(t)$ is reflected in the increase in the value of existing collaterals such as housing and structures at equilibrium states.

The business entities that can engage in the lending process to create the derived money are under the control of regulations and laws in the legal structure through the political structure. The lending-borrowing activities in the debt market through the fractional banking create environments for rent-seeking and wealth transfers through debt manipulations and interest charges, without real production and the creation of real wealth in the real sector. The size of the rent-seeking environment will depend on the nature of the extensive and intensive regulatory regime in the financial market. From the established rent-seeking environment, the initial secondary money from the debt market becomes the financial output where the maximum debt production as allowed by the law is equal to

$$F(t) = F(E(t), B(t), H(t)) = DE^{\alpha_1} H^{\alpha_2} B^{\alpha_3} = k\mathbb{B}(t), \qquad (4.53)$$

The exponents $\alpha_1, \alpha_2, \alpha_3 \in [0,1)$, such that $\sum_{i=1}^{3} \alpha_i = 1$ are debt distributive parameters of various categories of debt. The total of the financial claims over commodities and wealth at any production time is $(1+k)\mathbb{B}(t) = \mathbb{D}(t)$. At this maximum financial output no secondary money $k\mathbb{B}(t)$ can be created by debt issuance. In this respect, the supply of debt is fixed. The legal conditions of the political economy are such that the primary money $\mathbb{B}(t)$ in the money market cannot be traded in the sense of buying and selling. However, the derived money $k\mathbb{B}(t)$ in the debt market can be traded as financial commodities through ingenious schemes called by various names such as financial engineering, where different categories of debt with coupon rates are manufactured, priced, marketed and sold at various prices.

The movements of these financial commodities are governed by the information structure and preferences of the buyers and sellers for capital gains depending on the various movements of the coupon prices. In this way, each debt instrument in the debt market becomes commoditized with a phantom value that is sustained by a belief system and related to a belief that there is some real wealth in the real sector that constitutes collateral of the debt instrument. It may, however, be pointed out that debt creation has benefit-cost implication for the political economy. However, if the total social cost outweighs the total social benefit, then the system will experience some instability depending on the size of net cost. The political economy obtains social benefits to the extent to which the debt creation enhances real wealth creation in the real sector, but not necessary wealth transfers to enhance profits and the earnings of the financial bureaucratic capitalist class. Such benefits must be seen relative to the social costs in the debt creation.

4.4.1.1 Debt Market as a Game Space for Commoditized Debt Instruments in Rent-Seeking

Rent-seeking to enhance profits through wealth transfers is an integral part of the financial sector through money gaming. The most activities in debt creation through rent-seeking are mainly real-asset-ownership transfers rather than real-asset creations, which technically, are opened to all. The complex nature of the game and the presence of information asymmetry have given rise to principal-agent services for fees in the financial sector. The nature and size of the rent-seeking in the financial sector will depend on how the regulatory regime connects the financial sector to the real sector. One way is to connect the money market to the real sector where the high powered money, as the primary debt instrument, is made to depend on the growth of real output and wealth. In this way, the $\mathbf{m} = \frac{\mathbb{B}(t)}{Q(t)}$ and $\mathbf{b} = \frac{\dot{\mathbb{B}}(t)}{\mathbb{B}(t)}$ in equation (4.52) are controlled. From such a specification, the growth of the value of loans will be constrained by the activities in the real sector even though this will not do away with phantom wealth that is mandated by fractional banking and the built-in systemic inflation and risk. Another way is to link the real sector to the cost of lending-borrowing activities through institutional controls in the debt market, thus constraining behavior and debt creation. Yet another way is to tighten the regulatory conditions that allow businesses such as depository and insurance agencies to belong to the club of lending as well as the kinds of lending activities that are allowed in the debt market. The results of the presence of rent-seeking activities in the financial sector and the creation of a phantom wealth through derived debt are to increase the concentration of wealth and income in the hands of few, in terms of profit and capital gains and in ownerships that restructure financial claims over real wealth.

The increase in concentration of wealth and income is to constrain the general purchasing power and reduce the demand for consumer goods and housing that can be serviced from wage income, since the excess of consumer goods and housing are held in profits and capital gains that are represented by the financial instruments. This creates imbalance between effective demand and supply of goods and services leading to increasing inventory accumulation. The supply-demand imbalance and the increasing of inventories in the real sector are then corrected by advancing more credit from profit and phantom wealth to the consumers at high rates of interest to increase consumer demand for consumer durables and housing, and sometimes the basic necessities. The process leads to increases in labor indebtedness. The increase in labor indebtedness leads to further transfers of claims from real income to the bureaucratic capitalist class in the financial sector. The advancing of credit increases the consumer indebtedness as shown by indexes $e(t)$ with growth $\dot{e}(t)$ and $h(t)$ with growth $\dot{h}(t)$ in Figure 4.2.

The borrowing-lending activities in the debt market are the easy way of enhancing the profits and earnings of the bureaucratic capitalist class through transfers of financial claims and changes in ownership of real wealth. The high profit of the process is such that the producers in the real sector find creative ways to participate in the borrowing-lending activities in the financial sector through other means, including amplification of the credit-card system. The credit-card system is a backdoor approach to create debt as secondary money that has claim over real commodities. The process of amplification of the credit-card system of supply-demand transactions has become so engrained in the political economy that other institutions of education and non-profit activities are intimately involved. It is the easiest way of enhancing revenues and profits without real production. The environment of the system of rates of interest and penalties is complex, and will increase in intensity in favor of profiteers and against the borrowers as these fees are under the controls of the lending institutions.

The choice variables in the debt market E, H and B, their rates of change $(e, h$ and $b)$ and their acceleration rates $\left(\dot{e}, \dot{h}(t)$ and $\dot{b}\right)$ are under the control of the bureaucratic capitalist class in the financial business sector where the engineering of phantom wealth through debt creation is its powerful instrument. The choice variables in the real sector production K and L, their rates of growth

$(k$ and $l)$ with their acceleration rates $(\dot{k}$ and $\dot{l})$ are under the control of the bureaucratic capitalist class in the real sector, where the production of goods and services is a powerful instrument in real wealth creation. The economic parameters in the real sector, and the financial parameters in the financial sector define the information structures which are under competitive struggle between consumers, labor as a class, and bureaucratic capitalist class in the real sector, and between the principal-agent class and the bureaucratic capitalist class in the financial sector under given political and legal environments that restrict behavior under the principle of fairness of trade and markets.

These political and legal environments are under the control of the government which can be used by the social decision-making core to create varying regulations and laws to alter the decision-choice behaviors in the established markets in both the financial and real sectors. The manner in which the political and legal environments are used by any of the continuous succession of the decision-making core determines the nature of the governance within the good-evil duality. We must add that in these class competitions in the Schumpeterian political economy, the bureaucratic capitalist class holds the ultimate power. It has succeeded to arrest power from the decision-making core and entrench itself as the law makers, thus effectively controlling the economic structure and demolishing the power of labor in the market system. The effective direct and indirect controls of the political structure allow the bureaucratic capitalist class to work with the process of legislation to weaken the intensive and extensive regulatory regimes through various actions on the deregulation. The action on the part of the bureaucratic financial capitalist class is to widen the space of rent seeking in the financial sector by changing the rules of the trading of debt instruments for rent-creation, rent-protection, rent-harvesting and debt payments.

The essential points to note are that deregulation in the financial market, particularly, the debt market, 1) increases rent-seeking activities that are non-real-wealth-creating, 2) increases income and wealth disparities in favor of the rich, 3) reduces general consumer demand for commodities and housing that can be serviced from wage income and savings, 4) increases the demand for consumer loans at high interests and, 5) widens the framework that the consumer loans are serviced from the phantom wealth and profits, which further increase income-wealth disparities by impoverishing the consumers through high interest payments and penalties which then create a chaotic feedback process. The financial sector becomes self-exiting through financial games. It becomes disconnected from the real sector without a self-corrective mechanism. This self-excitement and non-self-corrective mechanism is fuelled by expectations of profit and rent-seeking exuberance under its own rationality and by constantly changing the δ-parameter for stability under conditions of unstable expectation. Deregulations mean that we do away with or reduce the ε-parameter in which case the $\dot{\mathscr{B}}(t)$ is not controlled or less controlled. To avoid the chaotic financial process, the (ε) must be regulated through $\mathbf{m} = \frac{\mathbf{B}(t)}{Q(t)}$ and $\mathbf{b} = \frac{\dot{\mathbf{B}}(t)}{\mathbf{B}(t)}$ to control the growth of phantom wealth, $\left(\dot{\mathscr{B}}\right)$ within limits that can be sustained by the real sector. In other words, the ε-parameter must be constantly adjusted by the government in the regulatory regime to the changing δ-parameter that is provided by the financial-sector activities.

Under these circumstances, the goal of regulation in the financial sector by the government or the decision-making core is to guard against the emergence of such chaotic conditions from uncontrollable money-game exuberance, as well as guide the real-financial system to stay on the stability path. This requires instituting a regulatory regime that will reduce over-production of financial instruments, manage the manner in which the commoditized debts are traded in the debt market, control an unnecessary advantage of principal-agent information asymmetry, curtail financial rent-seeking to a stable size, and ensure that the index of financial aggregate, as measured in real terms, is close to the index of real output (see Figure 4.6) through the regulation of its impact on the wealth of the nation. In this way, the phantom wealth is reduced to a manageable minimum to provide the system with a stability bubble, and the engineering of the systemic risk to the level that is sustainable by the real sector in the political economy. This is one way of looking at the engineering of social risk as discussed in [R7.16]. The point here is that the fractional banking with borrowing-lending activities has endogenized a financial bubble, phantom wealth and monetary game that have created systemic risk for the Schumpeterian political economy. The role of financial regulation is to control this systemic risk to a sustainable size by creating and implementing extensive and intensive regulatory regimes that will manage the financial bubble of the bubble economy.

4.4.2 The Interest Rate as a Connector of the Real and Financial Sectors through the Debt Market

The financial sector is basically dominated by the debt market where debt has been commoditized as *real something* for buying and selling in the financial sector. The concept and the process of buying on credit have transformed the concept of ownership rights. The consumer durables, such as automobiles bought on credit (loans) have joint ownership rights contained in the debt coupons. Houses bought on credit (loans) also have joint ownership inscribed on the debt coupons. Technically, these debt coupons are collateralized by future income and represented as something real. Price reduction in the future markets for these durables convert the real something into phantom wealth. Since the debt coupons contain phantom wealth that is not collateralized from the real sector, an engineering of a substitute must be found. The real sector *collateral principle* is now replaced by the *creditworthiness principle* on the basis of justified faith that is broadly defined to include such things as earnings capacity, equity, durables and others. In this way, the commoditized financial sector generates many differential outputs (financial product differentiation) for the debt market. The individual real wealth-no-real-wealth duality has estimated value on the basis of fair market value as collateral for a loan-no-loan duality. This fair market value projects a distribution of degrees of ability to pay between real wealth-no-real-wealth duality to which one can satisfy his or her debt obligation when a loan is granted. The creditworthiness-no-creditworthiness projects the degree to which one is capable of satisfying his or her debt obligation even if one has no real asset. The debt market is available and open to all public and private decision agents in the political economy. The interesting and most important thing associated with the credit worthiness principle is that the future output in the political economy has been used to collateralize loans of current consumption. The relational structure is diagrammatically shown in Figure (4.3).

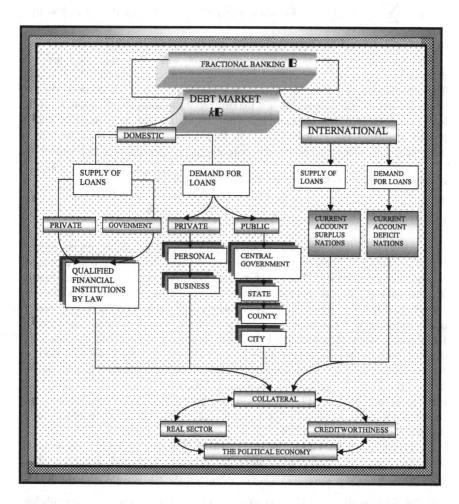

Fig. 4.3 The Geometry of the Debt Market and Participants in the Political Economy

The buying and selling of the debt instruments implies that we can define and specify the financial sector in terms of its commoditized outputs.

$$\mathscr{F}(t) = \left\{ y_j(t) = G_j\big(d_j(t), m_j(t)\big) \mid d \in \mathbb{D},\ m \in \mathbb{M} \text{ and } j \in \mathbb{J} \right\} \quad (4.54)$$

The meanings of $\mathbb{D}$ and $\mathbb{M}$ are as provided in definition (4.2).

The corresponding prices for the financial products are $\mathbb{P} = \left\{ \pi_j(t) \mid j \in \mathbb{J} \right\}$ and the total debt market value $\mathbf{V}$ is

$$\mathbf{V} = \sum_{j \in \mathbb{J}} \pi_j y_j \gtreqless (1+k)\mathbb{B}, \text{ where greater than implies capital gains, less than}$$

implies capital losses and equality implies neither capital gains nor losses.

The demand for loans is a decision to borrow while the supply of loans is a decision to lend on the basis of the interest rate given an associated risk of default. These two decisions can be formulated in terms of utility maximization on the part of demanders of the loan and profit maximization on the part of suppliers of the loan on the basis of the interest rate. The formulations of these decision-choice problems are not our concern here. On the basis of the framework here, it is postulated that when the optimization problems are formulated and solved we obtain sensitivity functions of demand for loans, $\mathbf{L}(r,Y)$, $\dfrac{\partial \mathbf{L}}{\partial r} \leq 0$, $\dfrac{\partial \mathbf{L}}{\partial Y} \leq 0$ and supply function of loans $\mathbf{S}(r,Y)$, $\dfrac{\partial \mathbf{S}}{\partial r} \geq 0$, $\dfrac{\partial \mathbf{S}}{\partial Y} \geq 0$. The demand for loans (debt) negatively depends on both aggregate income and aggregate interest rate. The supply of loans on the other hang positively depends on both aggregate income and interest rate. The two functions constitute a supply-demand duality. The demand for loan is inversely dependent on the rate of interest and income through savings. The supply of loans is positively dependent on the rate of interest and income through investment. As the cost of borrowing increases, the demand for loanable funds falls because it is a cost. Similarly, as income increases, the demand for loanable funds also falls since it reduces the willingness to borrow. Reverse relationships hold for the supply of loans through the savings process. As income increases, it presents an increasing ability to loan, hence the supply of loanable funds is positively related to income. Similarly, as the interest rate increases, it presents a willingness to increase the supply of loanable funds because the interest rate is a benefit.

The financial sector equilibrium is a market clearance of supply and demand for loans on the basis of aggregate interest rate and income where $\mathbf{L}(r,Y) = \mathbf{S}(r,Y)$, providing us with a loan equals debt condition as the $\mathbf{L} - \mathbf{S}$ equilibrium condition in the financial sector market system. The equilibrium condition in the real sector is the investment $\mathbf{I}(r,Y)$ equals savings $\mathbf{S}(r,Y)$ condition that provides us with the I-S equilibrium condition for the real sector market system. As it stands, we may solve the I-S equilibrium relation and $\mathbf{L-S}$ equilibrium relation to obtain the equilibrium condition for the political economy where the interest rate is negatively related to income, where $r_{\text{I-S}}(Y)$ with $\dfrac{dr_{\text{I-S}}}{dY}<0$ for the I-S condition, and $r_{\mathbf{L\text{-}s}}(Y)$ with $\dfrac{dr_{\mathbf{L\text{-}s}}}{dY}>0$ for the $\mathbf{L-S}$ condition. The possible geometric structure is shown in Figure 4.4.

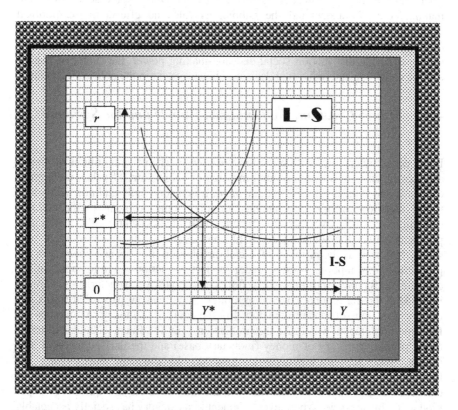

Fig. 4.4 The Geometry of Real-Financial Sector Equilibrium

4.5 Cost-Benefit Rationality, The Fuzzy-Financial Bubbles, The Optimal Bubble and the Management of the Systemic Risk

We shall now examine the nature of the phenomenon of the financial bubble and the strategies for managing the associated systemic risk by policy actions on the size of the financial bubble in the political economy. We shall then derive the conditions under which a political economy becomes a bubbled economy supported by a bubbled politico-legal structure to become a bubbled political economy. From definition (4.2), a financial bubble $\mathscr{B}$ is said to exist in the political economy if the financial output is partly collateralized from the real sector. Any non-collateralized financial instrument is just like printing money without the production of output in the real sector. In this case the financial bubble is positive when the value of the financial production exceeds the value of available goods and services. In this respect, the nature of the financial bubble may be seen as a measure of the relational value difference between the values of productions in the financial and real sectors. The financial bubble is the result of fractional banking as mandated from the legal structure by the social decision-making core. To understand the bubble, its role and effects in the political economy, we must examine the relational structures of production, income distribution, inventories and loans, all in real values. It is also useful to relate these structures to the distribution social decision-making power, the center of control and the collective and individual decision-choice system in the political economy.

4.5.1 Money Creation, the Financial Bubble and the Systemic Risk

Every economy operates with some form of money. The general utilities of money are seen in terms of unit of account, measure of value, store of value, medium of exchange and standard of deferred payment. It is theoretically and practically safe to say that these utilities are intended to solve some problems in the political economy of collective actions where individuals enter with differential preferences over the social cost-benefit configuration. The major problem may simply be stated as the problem of social distribution of costs and benefits in the production-consumption system under collective actions. From these utility characteristics, any instrument that fulfils these utility characteristics may be included in the definitional concept of money. Any financial paper that fulfils any of these functions of money is included in the concept of money. All monies are financial instruments which are the production from the financial sector in terms of the primary category of money (monetary base) and the derived categories of money (financial instruments). In general, the financial instruments are transformable to one another and to basic money which may be seen in terms of categorial conversion. The financial instruments constitute a sequence of paper production in categories where every element in a category has socially acceptable value; and category constitutes either a derived or a primary category to one another. The convertibility of categories is possible under actual-potential duality where the potential may be conceived as naturally unlimited while the actual lives in a socially bounded domain. The presence of money allows an easy commodity distribution from production output among the members of the social set-up since the money obtained can be used to hold inventories or to acquire consumer goods for consumption.

The presence of money is to allow an easier and practical solution to the problem of cost-benefit distribution among production-consumption agents in the political economy. In this respect, the political economy may be cast in two configurations of real cost-benefit configuration and financial cost-benefit configuration that must be related in terms of real goods and services on one hand and financial instruments on the other hand. Let us refer to Figures (3.2), (3.4) and (3.7) in terms of the relational structure of the real and financial sectors. Some additional income-wealth accounting is needed for the extension of equation (3.2) of Figure 3.8. An argument has been made here that the value of the real-sector production is the collateral for the value of the financial-sector production at any moment of time in the political economy. Furthermore, national loans are advanced in the political economy on the basis of real accumulated savings where the payments of such real loans are guaranteed by future real income of the borrowers. The lending-borrowing process functions in an increasing complexity depending on the regime of income-wealth distribution that is accepted in the political economy. The nature of income-wealth distribution is a reflection of the nature of cost-benefit distribution. The additional income-wealth accounting and the resulting equations involve aggregate inventory and its value in the real sector. Generally, the real sector may be seen in two. Let us keep in mind that in the real sector, aggregate savings are accumulated real commodities and services which are registered as accumulated inventories. In the financial sector aggregate savings are accumulated financial instruments that are registered as accumulated claims. At any moment of time two accounts may be specified for the real sector.

$$
\left.\begin{array}{ll}
Q(t) = C_1(t) + \mathbb{S}(t) & \text{a) consumtion-savings identity} \\[2mm]
Q(t) = C_2(t) + \mathbb{I}(t) & \text{b) consumption-investment identity} \\[2mm]
C_2(t) - C_1(t) = \mathbb{S}(t) - \mathbb{I}(t) & \text{c) production syncronization condition}
\end{array}\right\} \quad (4.55)
$$

In eqn. (4.55) $Q(t)$ is current output, $C_1(t)$ is the consumption goods and services, $\mathbb{S}(t)$ is the unused goods and services, $C_2(t)$ is the production of consumption goods and services and $\mathbb{I}(t)$ is production of investment and related goods. Perfect production synchronization requires that $C_2(t) = C_1(t)$ in which case the system is such that the production of consumer goods matches the demand for consumer goods. Similarly, the production of investment goods is exactly matched by the demand of investment goods. The system is producing exactly what consumers need and what investors need.

The conditions specified in eqn. (4.55) are useful in examining real finance in terms of additions to accumulated wealth that can constitute collateral in support of the financial-sector production of financial claims for the available past and present real wealth. This collateral support at any time point is the current potential supply $\mathbb{Z}(t)$ which is made up of accumulated inventory at that point, $\mathbb{V}(t)$ and current output $Q(t)$ that allow us to write current supply as:

$$\mathbb{Z}(t) = \mathbb{V}(t) + Q(t) \quad \text{supply conditions} \qquad (4.56a)$$

The total output constitutes the factor income which is divided between total wages, $W(t)$ and total profit $\Pi(t)$ computed in real terms such that $Q(t) = W(t) + \Pi(t)$. Given such income distribution, the total supply as specified in eqn.(4.6.1.2) may be written in terms of factor income and accumulated inventories. Thus,

$$\mathbb{Z}(t) = \mathbb{V}(t) + W(t) + \Pi(t) \quad \text{supply conditions} \qquad (4.56b)$$

From eqn.(4.56b) we can specify total available income-wealth savings $\mathbb{S}_T(t)$ as

$$\mathbb{S}_T(t) = s_w W(t) + s_\Pi \Pi(t) + \mathbb{V}(t) \quad \text{savings condition} \qquad (4.56c)$$

The accumulated inventories $\mathbb{V}(t)$ is the same thing as the accumulated real savings and hence eqn.(4.56c) is the current value for the lending-borrowing process. Let us suppose that these real values are structured in a constant aggregate price $P(t)$ that the monetary base $\mathbb{B}(t)$ is consistent with each production output, and under fractional banking $(k\mathbb{B})$ is the maximum debt creation where $(1+k)\mathbb{B}(t) = \mathbb{D}(t)$ constitutes financial claims over $\mathbb{S}_T(t)$. The financial bubble at any time period may be specified as:

$$\mathscr{B}(t) = \begin{cases} \mathbb{D}(t) - P(t)\mathbb{S}_T(t) \\ \text{or} \\ (1+k)\mathbb{B}(t) - P(t)\mathbb{S}_T(t) \end{cases} \qquad (4.57)$$

The size of the financial bubble defines the condition of systemic risk as well as indicates the possibility and the probability of the system's failure. The nature of the systemic risk depends on the amplification of the debt instruments, the buying and selling of the derived categories of the financial instruments, the expectation and speculative behavior around the buying and selling prices of the elements in the various derived categories of debt, the rate of turnover of the financial instruments and the rate of inflation of the collateral elements in the real sector. In monetary terms there are two types of prices that are relevant to the nature of the bubble. They are the aggregate price in the real sector and the aggregate price in the financial sector. The expectation and speculative behavior in the financial sector affect the aggregate value of derived debt where inflation enlarges the value of derived debt and hence the aggregate debt from $\left(k\mathbb{B}\right)$ to $\left(k\hat{\mathbb{B}}\right)$, where $\left(k\mathbb{B}\right) > \left(k\hat{\mathbb{B}}\right)$. Deflation performs the opposite. For the system's stability the aggregate price of the collateral must move in the same direction to maintain some notion of parity of monetary values in the two sectors. One of the important functions of fractional banking and the corresponding debt creation is the management of transfers of wealth ownerships of aggregate inventory, which is the real aggregate savings, to smooth out the relational structure between the surplus and deficit spenders as a result of income distribution, and individual behavior and preferences.

For sustainable systemic risk and the system's stability, there must be an optimal bubble with a stability behavior around it. In this respect, the management of such a stability behavior is under the activities of the decision-making core through the rules and regulations. These rules and regulations must affect the policy parameter (k) in creating the various categories of *derived debt,* as well as in controlling the number (n) of types of allowable categories of financial products in the financial sector of the political economy. The systemic risk, $\mathscr{R}$ and the bubble $\mathscr{B}$ depend on (k,n), the debt-creation parameter and the number of categories of financial instruments. While in general (k) is controlled, (n) may or may not be controlled. If (n) is not controlled then it is randomly determined by market forces using fuzzy information.

The parameters (k,n) must be related to $(\varepsilon - \delta)$ stability conditions as we have discussed above. The values (δ, n), are market-behavioral parameters and the values (ε, k) are policy-controlled parameters under the social decision-making core. The value of the parameter k is set by the social decision-making core and (δ, n) parameters are created by the transactors' collective behavior in the market under fuzzy-stochastic information structure which is then amplified by deceptive information structure. The parameter ε is then selected to control the system's stability dynamics. The parameter k is related the real output through the monetary base while the parameter ε relates to the overall dynamic real-financial behavior of the political economy.

4.5.2 The Optimal Financial Bubble, System Risk and Fuzzy Rationality

The question that arises is how the optimal financial bubble can be determined. Given the technique and method of determining the optimality condition and the optimal value of the financial bubble, can it be said that the optimal value is the same or different for different political economies? For any given political economy, the determination of the optimal financial bubble must be related to cost benefit balances of the role that the bubble performs in the management of the system. The financial bubble must be seen as a fuzzy-random variable due to speculative behavior and expectation formation with fuzzy-stochastic information that account for vague information and limited information wrapped in possibility and probability of outcomes. The expected bubble in the system at any time point must be computed in a fuzzy-stochastic space by using fuzzy-probability values. Let the aggregate expected bubble:

$$\hat{\mathscr{B}} = \mathrm{E}\left(\mathscr{B}\right) = \mathbf{P}\mathscr{B}\left(t\right) = \mathbf{P}\left(\left(1+k\right)\hat{\mathbf{B}}(t)\right) - \mathrm{P}(t)\mathbb{S}_{\mathrm{T}}\left(t\right) \qquad (4.58)$$

In eqn.(4.58) the real sector is assumed to be speculation free for analytical convenience, the value $\mathbf{P}$ is a fuzzy probability or probability of a fuzzy event and hence $\mathrm{E}\left(\mathscr{B}\right)$ is a fuzzy expected bubble [R7.15][R7.16] [R6.54]. The bubble as a decision outcome is a cost-benefit phenomenon that relates to the fractional banking decision. The bubble has a set of characteristics $\mathbb{X}$ which may be partitioned into a fuzzy benefit characteristics set, $\mathbb{B} \neq \varnothing$ and fuzzy cost characteristics set $\mathbb{C} \neq \varnothing$ such that $\mathbb{X} = \mathbb{B} \cup \mathbb{C}$, and $\left(\mathbb{B} \cap \mathbb{C}\right) \neq \varnothing$. The corresponding membership characteristic functions of benefit and cost are as specified in eqn.(4.59).

$$\begin{cases} \mu_{\mathbb{B}}\left(\hat{\mathscr{B}}\right), \dfrac{d\mu_{\mathbb{B}}\left(\hat{\mathscr{B}}\right)}{d\hat{\mathscr{B}}} \leq 0, & \left(\text{benefit characteristics function}\right) \\[4mm] \mu_{\mathbb{C}}\left(\hat{\mathscr{B}}\right), \dfrac{d\mu_{\mathbb{C}}\left(\hat{\mathscr{B}}\right)}{d\hat{\mathscr{B}}} \geq 0, & \left(\text{cost characteristics function}\right) \end{cases} \qquad (4.59)$$

There are economic and management reasons for the shapes and the slopes of these membership functions. Let us keep in mind the meaning of the nominal value of savings as the nominal value of inventories that constitute the collateral for financial output where the value of the bubble is non-collateralized. At the low value of the bubble, the social benefit of the bubble to the social management of inventories (savings) is high while the social cost of the bubble is very low. As the bubble increases, a value is reached where the social benefit begins to fall to reduce the fuzzy benefit characteristic set. Similarly, the social cost of the increasing bubble begins to rise after a certain value as the fuzzy characteristic set increases. The reason behind this is found in the logic of duality and fuzzy continuum where every benefit has a cost support and every cost has a benefit support, forming the decision unity under fuzzy-stochastic information that generates inexact probability and fuzzy-stochastic risk. To obtain an optimal bubble for the political economy, a fuzzy-stochastic decision problem may be formulated from eqn.(4.59) where we keep in mind that the bubble is a fuzzy-random variable [R6.41] [R6.54][R6.63][R6.101][R7.14]. The fuzzy-stochastic decision problem $\Delta = \mathbb{B} \cap \mathbb{C}$ with a membership characteristics function is defined as $\mu_\Delta(\mathscr{B}) = \left[\mu_\mathbb{B}(\mathscr{B}) \wedge \mu_\mathbb{C}(\mathscr{B})\right]$. The problem is to optimize the fuzzy membership function $\mu_\Delta(\mathscr{B}) = \left[\mu_\mathbb{B}(\mathscr{B}) \wedge \mu_\mathbb{C}(\mathscr{B})\right]$ as a fuzzy mathematical programming problem. The problem is equivalent to optimizing the fuzzy decision function $\mu_\Delta(\hat{\mathscr{B}})$ where the benefit-cost characteristics are internalized by the decision-making core through the membership characteristic functions as part of the management of the real-financial stability of the political economy.

Theorem 4.1

The optimization of the fuzzy decision problem $\mu_\Delta(\mathscr{B}) = \left[\mu_\mathbb{B}(\mathscr{B}) \wedge \mu_\mathbb{C}(\mathscr{B})\right]$ is equivalent to

$$\underset{\hat{\mathscr{B}}}{\mathrm{Opt}}\, \mu_\Delta(\hat{\mathscr{B}}) = \begin{cases} \underset{\hat{\mathscr{B}}}{\min}\, \mu_\mathbb{C}(\hat{\mathscr{B}}) \\ \mathrm{s.t}\quad \left[\mu_\mathbb{B}(\hat{\mathscr{B}}) - \mu_\mathbb{C}(\hat{\mathscr{B}})\right] \le 0 \end{cases} \qquad (4.60)$$

The proof of this theorem may be found in [R3.1.23] [R6.75] [R7] [R7.43] [R7.52]. The analytical geometry of the solution of this fuzzy-decision problem under soft computing is provided in Figure (4.5). In obtaining the numerical solution, the membership characteristics function must be carefully selected where the numerical definition of the bubble $\mathscr{B}$ and the corresponding fuzzy probability distribution are related to the definitional variables in the real and financial sectors, to connect the aggregate value of collateral to the aggregate value of the financial asset. The top figure shows cost-benefit duality with a continuum where every benefit has a cost support and every cost has a benefit support. The cost-benefit duality is then related to the systemic risk of the political economy. The bottom figure is a mathematical description of the top figure and relates to the nature of the fuzzy decision problem and its solution to obtain the optimal bubble $\hat{\mathscr{B}}^*$ with fuzzy-stochastic conditionality $\mu_\Delta(\hat{\mathscr{B}})$ that is related to the degree of acceptance of the systemic risk.

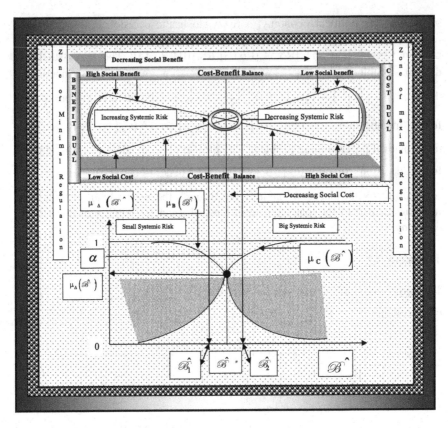

Fig. 4.5 An Analytic geometry of Soft-Computable Bubble, Cost-Benefit Duality in a Continuum and in Relation to the Fuzzy Laws of Thought and Approximate Reasoning

The solution set $\left(\hat{\mathscr{B}}^*, \mu_\Delta\left(\hat{\mathscr{B}}^*\right)\right)$ will depend on the mathematical description of the fuzzy-stochastic decision structure of the objective and constraint. Two examples of mathematical specification of the needed membership characteristic functions are suggested in Figures (4.6 – 4.9). The values a_1 in Figures (4.6) and (4.7) and a_1 in Figures (4.8) and (4.9) may be taken to be equal to the monetary base to the extent to which it is created to be consistent with output in the real sector. The system is risk-free when $\hat{\mathscr{B}} \in [0, a_1]$ and $\hat{\mathscr{B}} \in [0, a_1]$ respectively.

$$\mu_{\mathbb{C}}\left(\hat{\mathscr{B}}\right)\begin{cases}=0 & , \quad \hat{\mathscr{B}} \in \left[0, a_1\right] \\ =\dfrac{\hat{\mathscr{B}} - a_1}{a_2 - a_1} & , \quad \hat{\mathscr{B}} \in \left[a_1, a_2\right] \\ =1 & , \quad \hat{\mathscr{B}} \in \left(a_2, \infty\right]\end{cases}$$

$$\mu_{\mathbb{B}}\left(\hat{\mathscr{B}}\right)\begin{cases}1 & , \quad \hat{\mathscr{B}} \in \left[0, a_1\right] \\ =\dfrac{\hat{\mathscr{B}} - a_1}{a_2 - a_1}, & \hat{\mathscr{B}} \in \left[a_1, a_2\right] \\ =0 , & \hat{\mathscr{B}} \in \left[a_2, \infty\right]\end{cases}$$

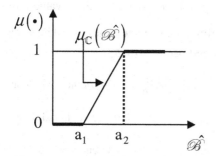

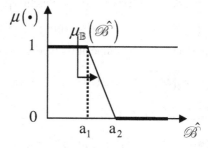

Fig. 4.6 Geometry of the Z-Fuzzy Number For "$\mathbb{C}$ is Big"

Fig. 4.7 Geometry of the Z-Fuzzy number for "$\mathbb{B}$ is Big"

$$\mu_{\mathbb{B}}\left(\hat{\mathscr{B}}\right)\begin{cases}=1 , & \hat{\mathscr{B}} \in \left[0, a\right) \\ =\dfrac{1}{2}\left\{1 - \sin\dfrac{\pi}{b-a}\left[\hat{\mathscr{B}} - \dfrac{1}{2}(a+b)\right]\right\} , & \hat{\mathscr{B}} \in \left[a, b\right] \\ =0 , & \hat{\mathscr{B}} \in \left(b, \infty\right)\end{cases}$$

$$\mu_{\mathbb{C}}\left(\hat{\mathscr{B}}\right)\begin{cases}=0 , & \hat{\mathscr{B}} \in \left[0, a\right] \\ =\dfrac{k\left(\hat{\mathscr{B}} - a\right)^2}{1 + k\left(\hat{\mathscr{B}} - a\right)^2} , & \hat{\mathscr{B}} \in \left(a, \infty\right]\end{cases}$$

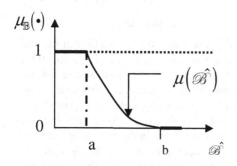

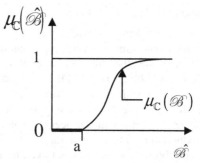

Fig. 4.8 Geometry of a Z-Fuzzy Number for $\mathbb{B}$ is big

Fig. 4.9 Geometry of a Z-Fuzzy Number for $\mathbb{C}$ is big

From the solution set and in addition to fuzzy decomposition, we can specify conditions of zones of systemic riskiness. Let the general policy preference define a degree of bubble preference associated with the systemic risk $\mu_\Delta \left(\mathscr{B} \right) = \alpha$. Associated with $\mu_\Delta \left(\mathscr{B} \right) = \alpha$, are two values of $\mathscr{B}_1$ with the benefit membership characteristic function $\mu_B \left(\mathscr{B}_1 \right) = \alpha$ and $\mathscr{B}_2$ that is associated with cost membership characteristic function $\mu_C \left(\mathscr{B}_2 \right) = \alpha$. With the value of $\mu_\Delta \left(\mathscr{B} \right) = \alpha$ we can develop a fuzzy partitioning of three zones of riskiness by using the fuzzy decomposition theorem [R6.55] [R6.63] [R6.75].

$$
\begin{cases}
\left\{ \mathbb{Z}_1 = \left\{ \left(\mathscr{B}, \mu_B \left(\hat{\mathscr{B}}^* \right) \right) \mid \mathscr{B} \in \left[0, \hat{\mathscr{B}_1} \right] \text{ and } \mu_B \left(\hat{\mathscr{B}} \right) \in \left[\alpha, 1 \right] \right\} \text{ small-systemic-risk Zone} \\
\left\{ \mathbb{Z}_2 = \left\{ \left(\mathscr{B}, \mu_\Delta \left(\hat{\mathscr{B}}^* \right) \right) \mid \mathscr{B} \in \left[\hat{\mathscr{B}_1}, \hat{\mathscr{B}_2} \right] \text{ and } \mu_\Delta \left(\hat{\mathscr{B}} \right) \in \left[\alpha, \alpha \right] \right\} \text{ manageable-risk Zone} \\
\left\{ \mathbb{Z}_3 = \left(\mathscr{B}, \mu_C \left(\hat{\mathscr{B}} \right) \right) \mid \mathscr{B} \in \left[\hat{\mathscr{B}_2}, \infty \right] \text{ and } \mu_C \left(\hat{\mathscr{B}} \right) \in \left[\alpha, 1 \right] \text{ large-systemic-risk Zone}
\end{cases}
$$

Zone one, $\mathbb{Z}_1$ describes a political economy with small risk due to the creation of financial instruments that have collateral support from the real sector. In other words, there is reasonable real-sector production in support of financial sector income-wealth transfers, thus the financial bubble is small. Zone two $\mathbb{Z}_2$ describes a case where the political economy is risky but manageable, in the sense that the production in the real sector is not enough to constitute reasonable collateral for the creation of debt instruments in the financial sector for the income-wealth transfers in the political economy. Here, the financial bubble must be under strict management to control the systemic risk through the aggregate parameters (k, n), and for real-financial stability through the aggregate control parameters (ε, δ). Zone three $\mathbb{Z}_3$ on the other hand describes a case where the political economy is extremely risky in the sense that the production in the real sector is substantially small in terms of absolute value and growth, to constitute collateral for the creation of debt instruments in the financial sector for the income-wealth transfers in the political economy. The size of the financial bubble is too large with too many financial claims over too few real output growth and wealth, creating an increasing systemic risk. Let us always keep in mind that the saving-borrowing process involves currently available collateral and the production of future collateral in terms of real income generation. In this case, the saving-borrowing process is linked by the time trinity of the past, the present and the future with the fuzzy-stochastic information structure for the decision-choice process, where lack of transparency amplifies the systemic risk as well as increases the rightward skewedness of income distribution.

4.6 A Note on Bureaucratic Capitalism, National Interest, Internationalism and Resource-Seeking

Let us now extend the decision logic of domestic rent-seeking to the conflict in the global resource constraint. The bureaucratic firms develop economic and financial empires in which the members of the bureaucratic class function as resource seekers to enhance profits, while the members of the decision-making core of the home government that represents their interest function as imperial officers in terms of diplomacy, the military and propaganda. This system can only function efficiently with the collaboration of the government's aggressive ideological plan and efficient propaganda machinery.

At the level of the general society, the national interest becomes completely or partially replaced by the interest of the bureaucratic capitalist class in its pursuit of more resources to enhance profits and wealth. Its global agenda in pursuit of resources and profits becomes the national imperial agenda which is then transformed into ideological beauty of globalization and development through trade and democracy. This ideological beauty is continually changing depending on changing competing interests of nations, global conflicts, and shifts of power relations in the global space of nations. The national goal-objective set in this situation is made to conform to the bureaucratic goal-objective set in support of the interests of the bureaucratic class. The interests of the bureaucratic class are seen in terms of profit visions which come to replace the social vision and national interest. The bureaucratic capitalism becomes transformed into bureaucratic socialism which then becomes transformed into bureaucratic imperialism that functions through the state and its power defined in terms of economic and military strength. Here, the bureaucratic financial and economic empires are promoted by the home state through the development of *crony imperialism* and neocolonial states whenever possible. The arsenals of the modus operandi include politico-economic destabilization of resource-rich and militarily weak nations, direct and indirect military actions, proxy wars under ideology of free trade without regard to fairness and equity in the global trading system. The concept and distorted practice of democracy are evoked without the understanding of the fact that the cultural confines of social systems with the systems' internal dynamics evolving, the disappearance of old institutions and the emergence of new institutions as different problems emerge in the social setup. This ideology of free trade is then backed by another ideology of democracy and individual freedom without regard to sovereignty, collective freedom and the national interest of the weak.

As the domestic rent-seeking environment becomes constrained by resource depletion and domestic social resistance increases, the domestic rent-seeking environment is transformed from a profit motive and fundamental ethical postulate of individual freedom to a global resource-seeking environment that defines an economic process of international rent-seeking activities. The domestic rent-seeking innovation and innovating class become a global resource-seeking innovation and innovating class respectively. The domestic rent-seeking innovation investment becomes a global resource-seeking innovation investment. The domestic rent-creation possibilities become transformed into global resource-creation possibilities. The domestic rent-harvesting also becomes transformed into global resource-harvesting. The domestic bureaucratic capitalist class is transformed into an international bureaucratic capitalist class. New international institutions of various forms are created by the bureaucratic class in addition to the existing ones, and used to enhance the validity of the ideologies in support of international invasions for rent-seeking and a resource-harvesting process. In a basic form, the bureaucratic domestic predators of bureaucratic capitalism become bureaucratic international predators in search of wealth and profits under bureaucratic imperialism and international imperial cronyism, sucking the global masses into the sphere of indirect slavery and sharp exploitation backed by military force of the governments they control as neocolonial states.

Democracy as a process of resolving conflicts of individual preferences in the collective decision space without violence is twisted into a meaningless ideology by the governments of bureaucratic capitalism, and used as an instrument to open the global resource space for international resource-seeking for rent-seeking activities without regard to costs to the environment and the inhabitants, and without a wealth creation for the country under predation. The ideological process of democracy is further intensified by another ideological support of free trade and the principle of free markets in the global politico-economic system, for the pursuance of financial phantom-wealth creation through currency speculation and a complex international credit system that the bureaucratic capitalists can create through their imperial governments. Democracy is no longer about a social decision-choice system for reconciling conflicts of individual preferences and will in the collective decision space, but rather democracy becomes an ideological instrument to manipulate collective preferences and will in the domestic and global spaces to conform to the preferences and will of the bureaucratic capitalist class. The bureaucratic capitalism is *morphologized* into a bureaucratic imperialism whose protection rests in the potential and actual uses of force and violence from the arsenals of war which is then backed by propaganda of mass deceptive information process. The interests of the bureaucratic capitalism become synonymous with the interests of bureaucratic imperialism which then establishes perpetual conflicts and war to justify perpetual production of technology of war, domestic controls and controls of resource-rich countries in the international political economy. The domestic rent-seeking games and the international resource-seeking games have payoffs that go to enrich the overall benefits of the members of the bureaucratic capitalist class where such benefits are directly or indirectly shared with the members of the social decision-making core. The behaviors and results of the two systems may be represented in a cognitive geometry as shown in Figure 4.10.

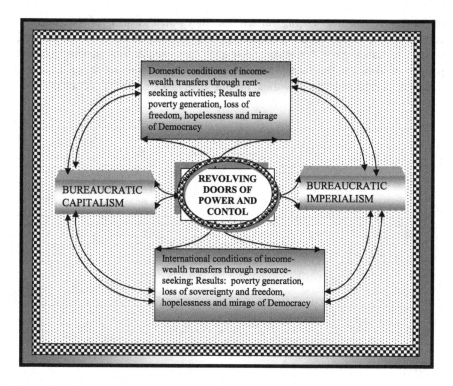

Fig. 4.10 An Epistemic Geometry of Dilemma of Democracy

The payoffs are not immediately defined in the quantitative space. They may initially be defined as qualitative variables in the qualitative space and then translated into quantitative variables in the quantitative space by appropriate quality-quantity mappings. The payoff values are part of the general incentive structure to motivate the choice of strategies to accomplish the control of domestic and international governments. Here, a dilemma of constitutionally elected representation emerges. The constitutional democracy and capitalism have the tendency to generate domestic conditions of rent-seeking activities in support of profit enhancement with a mirage of participation of the members of the principal, especially under bureaucratic capitalism and bureaucratic imperialism. At the level of domestic political economy, the process is such that the government of the people's representation is transformed into the government of the representation of the bureaucratic capitalist class. The people's capitalism and people's freedoms are transformed into bureaucratic capitalism and freedoms for bureaucratic capitalist class. The people's interests, instead of been represented are substantially neglected where labor power is substantially dismantled. The practice of democratic collective decision-choice system instead of serving the people to preserve their freedoms is transform to generate poverty, curtail freedoms and institute fear. The integration of domestic rent-seeking and international resource-seeking under the integrated works of bureaucratic capitalism and bureaucratic imperialism generates injustice, international poverty, wars, colonialism, neocolonialism and fear. It strips wealth from nations, robs the people of their bright future and enslaves labor in the name of freedom, democracy and development without the ability of labor's effective actions.

Chapter 5
Democracy, Political Markets, Socio-Political Responsibility and Accountability

5.1 Market Relations, Accountability and Social Formation

We now turn our attention to examine social responsibility and accountability of the decision-making core by relating rent-seeking and resource seeking in the democratic decision-making system. To relate the rent-seeking phenomenon to domestic socio-political accountability, we may explore some essential characteristics of the political process under two social regimes in a democratic decision-choice system: one with a political market and the other without for which accountability is to be examined. We shall also examine the resource-seeking phenomenon to international socio-political accountability and responsibility in the global socio-political system of nations. A political market has supply and demand interactions that may lead to political equilibrium or disequilibrium which may also be viewed in terms of cost-benefit duality. As a supply-demand process, the market must have identified product(s) that must relate to the needs and wants of suppliers and demanders at unspecified price(s). The political market of interest here is a market whose output is *influence* that is generated through access and whose incentive structure is market-defined in some form of a pricing system. This is consistent with the Schumpeterian political economy. In a non-political market however, there is an influence through access which operates differently through different social mechanisms whose incentive structure is moral principles and not through buying and selling mechanisms. This is consistent with the Marxian economy but may also operate in the Schumpeterian economy. In a democratic organization at the national level, accountability relates to national interest in creating the goal-objective set after the establishment of the decision-making core by the principle of majority rule.

A question arises, therefore, as to the extent to which the interests of constituencies and party affiliations relate to the overall national interest and social vision. Are the members of the decision-making core accountable to the interests of their constituencies or party affiliations or the interests of the nation or all of them? How is the social responsibility defined over the governed? The democratic decision-making process is said to be functioning efficiently if there are no political markets for the buying and selling of influence that goes to affect the national interest formation, the creation of the elements in the goal-objective set and the budget-revenue process for implementation. This does not mean that there is no social influence distribution as well as no effects on the setting of national goals. It simply means that the acquisition of access and use of influence from the general public in affecting the direction of the social vision and national interest with the supporting goal-objective formation are directed principally to the welfare of the nation and its people on the non-market principle in the best assessment of cost-benefit configuration.

In the regime of *non-political market,* decision-making by the members of the core is driven by higher moral principles of national interest and the peoples' welfare, that is, by collective welfare rather than by individual welfare or interest. In non-political markets the incentive structure and rewards to the members of the decision-making core are principally non-material while in the political market the incentive structure and reward is directly or indirectly material for the individual. The question, then, is what characteristics define the political markets. Are transactions in the political market part of the whole social-corruption structure to distort the efficiency of collective decision-making in the democratic social formation in order to favor private gains to the disadvantage of the general public? Are the political markets not institutional frameworks for transferring income and wealth from the have-nots to the haves?

Political markets are said to exist if access and influence are purchased in order to affect the setting of the social vision, national interest and the formation of the social goal-objective set for personal and private gains and benefits. The existence of political markets, as is being used here, is that there is selling and buying of access to the members of the decision-making core in order to influence the social decision-making process for private gains. The buying and selling may be direct or indirect and legally or non-legally accepted. The private gains take the form of rent creation, preservation and harvesting from the structures of the national interest and the supporting social goal-objective set as have been discussed in the previous chapters. In the regime of political markets, individual members may make decisions through their votes that are shaped by the activities in the political market through the supply-demand configurations. Access to the members of the decision-making core is not centrally democratized to the demands of the democratic decision-making system. The two social structures characterizing the democratic decision-making system may be presented as accountability and responsibility of the social relationships of the sovereignties of the citizens and the decision-making core in the presence of political and non-political markets in Figure 5.1.

We now turn our attention to examine the role of the political and non-political markets relative to socio-political accountability and responsibility of the sovereignties of the voting public and the members of the decision-making core in a democratic decision-choice system as it is represented in Figure 5.1. Let us keep in mind that political accountability and responsibility are related to governance and not the government. The governance is under the control of the decision-making core who must balance their preferences against the preferences of the public under majoritarian decision-choice system. The government has institutional neutrality and cannot be held to accountability and responsibility.

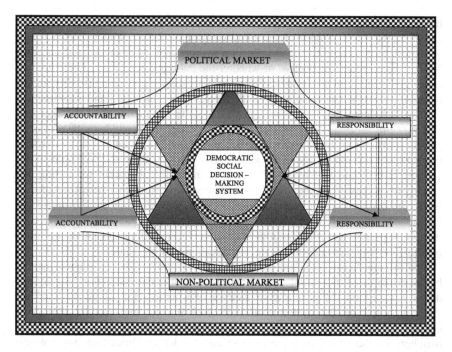

Fig. 5.1 Accountability-Responsibility Geometry in the Democratic Collective Decision-Choice System

5.1.1 The Nature of the Concept of Non-political Market in Democratic Decision-Choice Systems

Ideal democracy is one that has no political markets in that the guiding principle of decision making by the members of the decision-making core is the welfare of the state and its people as seen on the basis of majoritarian principles in the democratic decision-choice system. Here, access to the members of the social decision-making core is of democratic distribution with citizen sovereignty in the sense that individual members have free access but acting alone may have no effect on the setting of the national interest, social vision and formation of the supporting social goals and objectives. The national interest and the supporting goal-objective set as determined by the decision-making core are taken as given and individual members in the constituencies cannot affect their construction. There is, however, the postulate of perfect information in the electoral process in selecting the members who will constitute the decision-making core. The citizens can organize to influence the members of the decision-making core as to the setting of national interests, social vision and the needed social goals and objectives, and their implementation on the basis of the welfare of the state and its people, but not on the basis of private interests, welfare and gains. Here, the citizens' responsibility is to become an informed social decision-choice participant in the selection (voting) process. This responsibility includes obtaining appropriate information regarding the social choice elements, sharing of the information and analyzing the information to create a knowledge base for an efficient participation.

Access to the members of the decision-making core is democratized on equal basis of citizenship irrespective of whether the person votes or not. There are two cases to consider in terms of political accountability and responsibility in the political economy. One case is where the access is citizen-based and constituency-unrestricted. The other case is where the access to the members of the decision-making core is both citizen-based and constituency-restricted on the basis of perfect information. In both cases, pork-barreling and rent-seeking do not arise if the general criterion of social decision-making is based on the principle of the welfare of the state and its people. There is no political market for selling and buying of access from the members of the decision-making core (the elected body). In this respect, social decision-making must be distinguished from private decision-making even though both decisions are based on democratic principles of individual preferences and freedom to exercise these preferences under a defined information structure and the rules of the decision-making game.

The private decision-making is distinguished from the social decision-making by the principle of individual interest and welfare. It is the distinction between the decision based on the principle of welfare of a nation and its people, and the decision based on the principle of individual interest and welfare that gives content to the notions of the "service to country". The individual access to the members of the social decision-making core under conditions of non-political markets is an access to share information on decisions that will improve the welfare of the nation and its people. It is an access that works on the principle of *each for all and all for each*, a principle that sees the collective interest of society as prior to the private interest where the individual is beholden to the society and the society is accountable to the individual progress. The government becomes the government for the people in the true essence of democratic social formation where social decisions are constitutionally guided. Democracy in the political economy is thus, by the people, of the people and for the people. The political accountability and responsibility in this democratic social decision formation must be examined in relation to the welfare of the nation and its people and not simply in relation to the welfare of the individual or some individuals in the nation. Here, interest groups compete to improve the collective welfare of the nation on the basis of differences in social ideology, methods of governance or both. Socio-political responsibility and accountability rest on the people; is of the people, and for the people.

5.1.2 The Nature of the Concept of Political Market and the Democratic Decision-Choice System

When the access to the members of the social decision-making core is seen as an access to define national interest and shape the elements in the social goal-objective set in order to create conditions of influence for private benefits at the expense of the welfare of the nation and its people, then the democratic process alters its form. Democracy exists in name only and at a rhetorical level with deceptive practices on the basis of a deceptive information structure, in order to confuse the people regarding the nature of socio-political responsibility and accountability. The objective, here, is to generate rent, protect rent and abstract rent in the national decision-choice system for the benefit of few, and at a cost to most. Unfortunately, most of the currently known democratic decision-choice systems are of this nature. Access acquires the character of a commodity for buying and selling for the acquisition of influence. The solution to the conflicts in the individual preferences in the democratic collective decision-choice system is given market imputation under monetary weighting system. The commoditization of access generates political market in which indirect or direct commercial transactions take place. Direct transaction where access is exchanged for something is defined in contemporary times as *illegal political corruption*.

An indirect transaction in this political market where exchange takes place is called *kick-back* transactions or a revolving door of *benefit-sharing* transactions. The indirect transaction is *legalized political corruption* of the democratic collective decision-choice system. In the regime of political markets, people seek political office to acquire a membership into the decision-making core (sometimes referred to as the lawmaking body) not for the reason to make decisions to improve the welfare of the state and its people but to acquire the capacity to produce *commodity influence* through membership that allows them to commoditize their access. For this to work, they have to falsely advertize themselves and social positions to deceive and conquer votes from the voters of their constituencies. Lying about oneself, sometimes referred to as misrepresenting oneself is not a crime against the state and is only punishable by possible loss of social confidence and possible loss of votes. The outputs of the political production in the political markets are access to the decision-making core and the influence to shape social decisions that create environments for individual private benefits through the rent-seeking process.

The emergence of political markets in democratic social formations is the planting of the seed of destruction of any meaningful foundation of democracy and its practice as a mode of collective decision making without violence. The seed of self-destruction must be seen in terms of political accountability, responsibility and a rent-seeking enterprise as well as an enterprise of corruption. The rise of political markets is basically an emergence of conditions where the political process and the social power invested in the government may be hijacked away from the service of the interest of the nation and its people to the service of individual private accumulation and wealth building, especially in the favor of the bureaucratic capitalist class. Internationalization of this process also destroys any meaningful development of global democracy by creating conflicts and war. The relationship of the society to the three structures is illustrated in Figure 5.2.

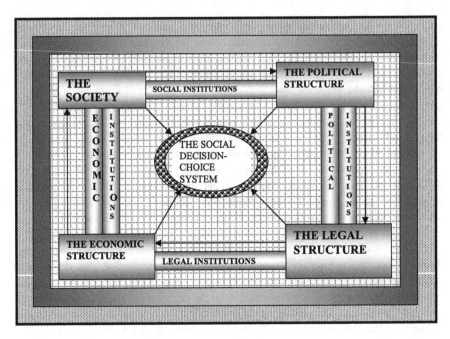

Fig. 5.2 The Analytical Geometry of Social Organization and Institutional Links and the Decision-Choice System

The destruction of the democratic relational structure through corruptive practices creates a possibility where the governmental power may be hijacked to serve interest groups with financial power to buy access and influence. At the level of domestic conditions, the hijacking of the power of state institutions may lead to the use of the state power to create injustices, human-rights abuses, civil-rights denials and an imposition of an unjust judicial system to extract rent without political responsibility and accountability. At the level of international conditions and depending on the military power of the state, the hijacking may lead to the establishment of a regime of perpetual wars by denying human rights of members of other nations, disregarding the interest of militarily weak nations and establishing the doctrine of the *right is in the bigger guns* in order to extract resources from other nations for privileged accumulation without political responsibility and accountability. This translates into the law of the jungle in the internationalism of lawlessness where *the bigger a nation's gun the bigger is the nation's right to be obeyed*. In this respect, there is a complete destruction of the stable relations among the economic structure, political structure and legal structure needed to establish democratic safeguards and check-and-balances against the possibility of human corruption, individual social abuses of the integrity of the collective social existence, and the use of the social power invested in government institutions to abuse citizens and create a decay of the human essence. The question now is what the conceptual meanings must be assigned to the political responsibility and accountability as they relate to the political, legal and economic structures of the nation. The relational structure of the political market to other important structures is shown in Figure 5.3.

The transformation of domestic bureaucratic capitalism to international bureaucratic imperialism where the bureaucratic capitalist class becomes the *bureaucratic imperialist class* is such that all countries are being exploited by the bureaucratic imperialist class to the full disadvantage of the laboring class. The governing class of the corporation starts to control the governments including the government of their home of residence for international resource-seeking and domestic rent-seeking for profit enhancements, where such profits are distributed to the members of the bureaucratic imperialist class for an ostentatious lifestyle. It will become clear that the current increasing corrupting influence in governance and economic management may be attributed to the rise of bureaucratic imperialism, the bureaucratic imperialist class with domestic rent-seeking and international resource-seeking activities in the global space where the driving incentive for decision is profit enhancement. In fact the labor in the home base of bureaucratic imperialist class under rent-seeking will be going through socio-economic power reduction at the bargaining condition just as the labor in the countries whose resources are under resource-seeking.

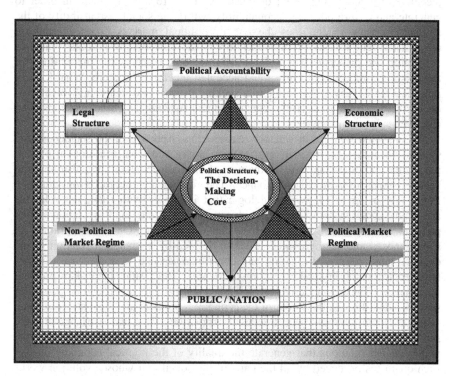

Fig. 5.3 Market Relations and Accountability in Democratic Social Formation

5.2   Political Domestic Accountability, Responsibility, Rent-Seeking Phenomenon and the Decision-Making Core

When we speak about rent-seeking in relation to political responsibility and accountability, a question arises about who are the members of the decision-making core accountable to under the electoral politics in any defined democratic collective decision-choice system. Are the members of the decision-making core accountable to the nation or to their constituencies or both or themselves and their cronies? A follow-up question arises as to the proportional distribution of total responsibility and accountability to the nation and the constituency for each member in the decision-making core if they are accountable to both. Similarly, to what extent do personal interests enter to shape the regime of socio-political responsibility and accountability and the quality and efficiency of the democratic collective decision-choice system? Are there conflicts of interests and if there are, to what extent do such conflicts of interest affect the rent-seeking environment, efficient governmental budget allocation and funding of elements in the goal-objective set? How is political accountability related to responsibility and incentive structure? What kind of incentive structure must be there in order to generate and ensure political accountability within the social responsibility of the nation? To what extent does the rise of political markets produce cronyism through the rent-seeking process and how is the rent-seeking process related to corruption and bribery? Rent-seeking activities take place within the confines of the existing politico-legal regime and accepted incentive structure. What kind of incentive regime gives rise to the rent-seeking process and maintains? What kind of incentive regime gives rise to corruption and sustains it? These questions need creative analyses and answers.

We are interested not in whether rent-seeking is present in the socioeconomic system, because it is, but why rent-seeking arises and what role does it play in intra-stage and inter-stage transformations of the socioeconomic organism through various stages of political governance. The relational structure is presented in Figure 5.4 in terms of pyramidal logic that is shown with triangular relations around the social system. One pyramid is presented as a triangular relation among political governance, accountability and corruption. On this pyramid is superimposed another pyramidal structure defined by a triangular relation of the decision-making core, the nation and constituencies and its logical relationship between accountability and corruption as the members assume the task of political governance. These interactive pyramidal relations evolve around a particular social system through the forces of the economic, political and legal structures in a synergetic configuration. (For a detailed exposition of the use of triangular logic see [R10.15] [R7.14]). The pyramidal logic allows us to affirm the conceptual notion of part-relations that generate the quality of the political economy, in that such quality is produced by all the parts of the social unit whose evolution works on the principles of opposites, conflicts, duality, polarity, unity, and continuum.

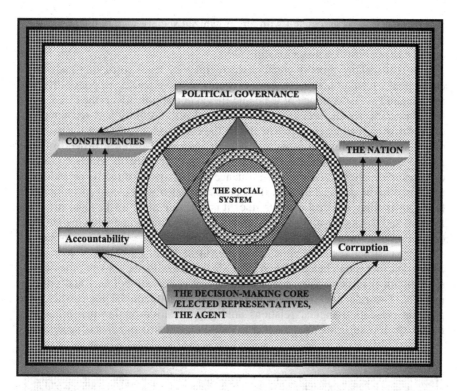

Fig. 5.4 The Pyramidal Geometry of Accountability-Corruption Structure of the Decision-Making Core in the Nation and Constituencies

It has been argued that rent-seeking, composed of rent-creation, rent-preservation and rent-harvesting, is intimately associated with a national interest definition and social goal-objective formation. It may also be associated with corruption and political accountability within socio-political responsibility on the part of the citizenry depending on how these are defined. The definition of national interest and the formation of a social goal-objective set are processes that create and establish environments of rent-seeking possibility, rules of governance, responsibility and accountability. The nature of the distribution of rent-seeking possibilities depends on the corresponding structure of distribution of social power to influence the members of the social decision-making core to establish the national interest and the goal-objective formation. It is also related to how such influence-tampering helps to define the legal structure at any legislative period to preserve the conditions of rent-seeking as well as to protect the interest of the rent-seeking class. The rent-seeking class is composed of rent-seeking innovation and non-innovation investors comparable to innovation and non-innovation entrepreneurial investors. The probability distribution to harvest rent after its creation depends on the regulation enforcement which in turn depends on the distribution of social power to influence the decision-making core regarding intensity of regulation. The rent-seeking phenomenon, like the phenomenon of social corruption, is associated with any organized society with political and legal structures that create laws and rules to regulate social behaviors in the social space on the basis of cost-benefit rationality.

The rent-seeking process creates two regimes of legal corruption and illegal corruption which affect the nature of political governance, responsibility and accountability. The presence of legal and illegal corruption affects the character of the democratic social formation, socio-political responsibility and accountability of the decision-making core. It also affects the collective character of the members of the society. Corruption is viewed here as the use of political office to create private benefits and public costs in favor of private benefits. Such corruption operates through an access and influence that may be directly or indirectly purchased in the political market. Bribery is associated with illegal corruption that functions in illicit political markets since it is not admissible by the legal structure. Indirect kick-backs, revolving doors, political contributions and lobbying are legal corruption that functions in the legal political market since they are permitted by some legal structures. The point is simply, these legalized practices corrupt the democratic decision-choice system. We shall deal with the two regimes of the socio-political responsibility and accountability under non-political and political markets. The relational structure is presented in a pyramidal logic where one pyramid presents triangular relations among political corruption that is legal or illegal. On it is superimposed another pyramid logical structure that is specified by

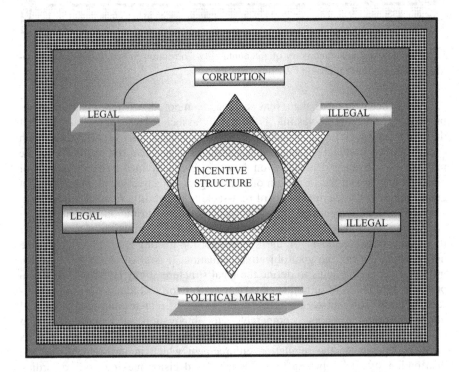

Fig. 5.5 The Relational Structure of Political Markets and Corruptions

a triangular relation among political markets, legal and illegal structures. The two logical pyramids revolve around the incentive structure. The relational structure of corruption is presented in Figure 5.5 where there are legal and illegal corruption and legal and illegal political markets that interact to affect the quality of the political economy under a democratic collective decision-choice system on electoral process.

5.2.1 Political Accountability and Responsibility under Conditions of Non-political Markets

In a democratic social formation, the relationship between the political and the legal structures as seen from the role of the decision-making core are such as to define boundaries of protection for the individual and the social entity in various combinations. The individual holds dual responsibilities which also translate to dual accountabilities. One responsibility is to self and the other responsibility is to the social collectivity. The individual and the social collectivity hold the title of the principal in any governmental system of governance. In a democratic decision-choice system the decision-making core assumes the responsibility of an agent of the collective and indirectly performs the social duties for all members in the collective. The role of the social decision-making core is to define and protect the collective interests of the nation and its people. The role of the constitution may be seen as defining the boundaries of private and public interests as well as individual and collective interests. The legal structure is designed to protect these boundaries and hence help to reconcile the conflicts in private-public duality as well as conflicts in individual-community duality. The political economy as we have defined in previous chapters is a plenum of forces of tension in individual and collective preferences where responsibilities and accountabilities have a tendency to alter forms. The responsibility of the social decision-making core is social decision-making and its implementation to create a control system in the legal structure, administering the social organization and managing its institutional set-up in a manner that produces the maximum social welfare of the nation and its people. The administering includes, among a multitude of things, social progress and stability, and the development of relevant supporting institutions to facilitate the smooth working mechanism of the democratic social formation on the principle of a democratic decision-choice system.

The criterion for examining political accountability must, thus, be framed around the notion that decision-making by the decision-making core in every instance is social cost-benefit driven but not private cost-benefit motivated on the basis of self-interest or interests of cronies. This is not the case in the management of the activities in the private sector where personal interests, viewed in terms of private cost-benefit balances, is the driving force of decision-making without considering the cost-benefit implications to the society. The social cost-benefit balances of the members of the decision-making core must in principle override their individual private cost-benefit calculations on the basis of their personal interests. This is their social commitment to serve the society and to administer the social responsibility of the government. Corruption tends to arise when the private cost-benefit calculations override the social cost-benefit calculations of the members of the decision-making core (the elected or non-elected body acting as the social agent on behalf of the populace who is placed as the principal). The test of social accountability of the members of the social decision-making core requires an *accountability measure* which will be a composite one given the defined social responsibility space. It consists of efficiency elements of all the social decisions relating to the functioning of state institutions and best practices in political administrations regarding the collective aspects of economics, politics and law. The specific elements entering into the construct of the accountability measure may vary from nation to nation, from generation to generation and their cultural practices given the constitutional mandates of the nations and their legal structures. They also have a time dimension for each nation as the social system evolves in quality and quantity.

In this framework, where political accountability that is associated with the members of the decision-making core is strictly social cost-benefit based, rent-seeking does not arise when the principle of social cost-benefit calculations is the foundation of decision-making by the members of the decision-making core. Private self-interest is submerged in favor of public interests. It is here that the concept of service to the nation acquires a special meaning and content because service by each member places the society's interests as prior to the personal interests. Conflicts, however, may arise in individual accountability and assessment of cost-benefit distributions among the interests of the nation, constituency of the individual representatives and their party affiliations. The preferences of the individual members of the decision-making core must be formed to reconcile nationalism, constituency affinity and party allegiance, given one's subjective scale of preferences. It is here that we must deal with the *nation-constituency-party* problem of accountability and the cost-benefit effects in social decision-making by the individual members in the social decision-making core. It is also here that pork-barreling tends to arise and the national budget becomes distorted away from the elements of the goal-objective set.

A question thus arises in the administration of the democratic collective decision-choice system. Is an individual member in the decision-making core accountable to his or her constituency's social cost-benefit bundle by electoral politics, accountable to the cost-benefit bundle of their party organization and ideology, or accountable to the national cost-benefit bundle and its fair distribution by the system's political organization as seen from the conditions over all national interests? In other words, the question of political accountability is a complex one under conditions of non-political markets. The relational structure that presents conflict zones of decision-choice actions is presented in Figure 5.6 in a pyramidal logical geometry where one pyramid presents itself as triangular relations among the interests of the nation, constituency and party. Superimposed on this is another pyramid that shows pyramidal relations of zones of conflict in preferences of members in the decision-making core in terms of cost-benefit balances of the nation, constituency and party. The geometry demands zonal analyses of conflicts in accountability, given the defined social responsibility within the culture of actions exemplified by a political duopoly.

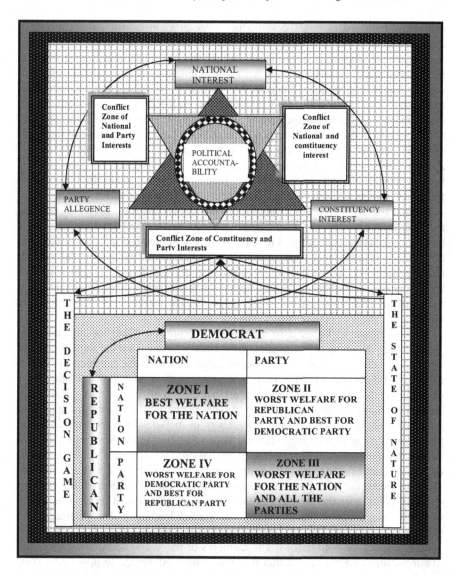

Fig. 5.6 Political Accountability, Political Interest and Conflict Zones in the Decision Game

The complexities of the problem of trilateral relations of a non-linear nature must be seen in two areas of the formation of the national interest and the supporting goals and objectives on one hand, and the general budget allocation decisions on the other. We have presented a structural analysis of the formation of the national interest and the supporting goal-objective set. The analysis was made on the assumption that the members of the decision-making core decide on the basis where their preferences are shaped solely by attaining the *best welfare* of the nation and its people. This criterion of the best national welfare condition overrides both party and constituency interests in terms of cost-benefit configurations. Thus accountability of the members in the national decision-making core is implicitly to the nation. When the national interest is established, the accountability is shifted to the act of designing a goal-objective set that will support the overall national interest. The formation of such a goal-objective set has been discussed in [R15.18]. The assumption in the decision-choice process in the formation of the goal-objective set is that the preferences of the members of the decision-making core are shaped by social cost-benefit conditions of the goal-objective set that will best support the national interest as determined. When the national interest and the supporting social goal-objective set are formed, they become constraints on decision-choice processes of the states in the case of federated states, constituencies, localities, cities, municipalities and politico-economic regions within the territorial boundaries of the nation. Given the national interest with a social vision and the established goal-objective set, budgetary decisions are tailored to support the goal-objective set after ranking. In this process of analytical work, utility may be substituted for cost-benefit where utility is measured in an arbitrary but consistent domain. Let us keep in mind that the theory of government budgetary allocation and process is the implementation theory of the goal-objective set in the support of the national interests and social vision.

With reference to Figure 5.6, the states of nature are either the nation or the party. In a political duopoly like that of the United States, both the Republican and Democratic Parties agree to operate in Zone I where the preferences of the members of the social decision-making core are in favor of national welfare. It is in this zone that bi-partisan politics are played in favor of the national interest. Here, everybody wins. Zone III is the case where a temporary impasse is established in the sense that all the members in the social decision-making core play complete partisan politics with the hope of obtaining the best net benefit for their respective parties at the expense of the nation. The welfare of the state and its members is not of a concern. The concern of each member is the winning strategy of the political party. In this zone the agent operates with a deceptive information structure that places the principal in the framework of information asymmetry. In so doing, the decision-choice actions are stalled leading to the destruction of popular confidence in the governmental decision-making structure and a loss of popular respect for both parties. The members of the social decision-making core (the legislature) soon find out that it is not to their advantage to operate in Zone III.

Party politics enters in terms of party ideologies by establishing the national interest and its vision. This interest and vision may involve the type and size of the government, the nature of society, the character and nature of law and economics, and a host of many important qualitative elements that will define the national spirit and shape the path of the social set-up and the social cost-benefit distribution. Party allegiance may come to cloud the social preference ordering of the members in such a way that the accountability to the national interest in terms of the welfare of the nation and its people is sacrificed in favor of the accountability to political party and its ideology to corrupt social interest in favor of some group interest. The nature of this party accountability in terms of allegiance is crafted within the internal dynamics of party organization and reinforced by a *carrot-and-stick* process in the decision-making core.

Similarly, a conflict may arise and does arise between national interest and constituency interest on one hand and between party interest and constituency interest on another, both of which may come to shape the decision-choice preferences and the accountability question of the members in the decision-making core as viewed in the cost-benefit space or cost-utility space. In these cases, the accountability of the members may be diverted from the national interest to the immediate interest of the constituency or from the constituency to the party interest. The individual member in the decision-making core, given his or her subjective preferences, must balance the possible conflicts among the three interests in addition to personal ones in terms of cost-utility computations. The individual knows that his or her return into the decision-making core depends on the perceptions of the members of his or her constituency given his or her party affiliation. These perceptions, in the last analysis, are translated into cost-benefit balances even if they are poorly assessed and subjectively calculated.

The question of political accountability is not simple and is further complicated by the organizational structure of the decision-making process of the core given the responsibility space. If the decision-making process is organized around committees and sub-committees, then the political accountability of the individual member will depend also on the committees to which one belongs, committee's rights to social decision-making, in addition to the degree of decision importance of the committee in the overall social decision-making process under a democratic collective decision-choice system. The underlying assumption is simply that all decisions in the social set-up work on the principle of majority rule. The whole decision-choice process takes place under a defective information structure composed of information limitation, vagueness and public ignorance that generate random and fuzzy outcomes in a fuzzy-stochastic space [R7.14] [R7.15][R15.14]. The information representation of the randomness and fuzziness of these outcomes are either fuzzy-random or random-fuzzy variables that capture the possibilistic-probabilistic uncertainties in the social decision-choice space. These uncertainties are amplified by the presence of a deceptive information structure composed of disinformation and misinformation characteristics in the social information set. All these define the need for cognitive tools of fuzziness, synergetics and complexity for their analysis and understanding.

5.2.2 Modeling the Political Accountability and Decision-Making Process through Cost-Benefit Rationality and Fuzzy Restriction

The establishment of committees and subcommittees is a partition of the social decision space into manageable parts. The quality of the decision-choice outcome will depend on the rules of aggregating the committees' decisions whether they overlap or not. The individual degrees of accountability to the three political segments may be viewed as fuzzy weights that can be fuzzy-tuned as the social environment tends to alter, given the individual preference ordering over the social space. The fuzzy structure and the fuzzy-tuning allow us to analyze the system's qualitative changes as the preferences of the members of the decision-making core shift in the zonal space. Let $A_i^n, A_i^c,$ and A_i^p be accountability-characteristic sets that relate to net cost-benefit balances $B^n, B^c,$ and B^p of the nation (n), constituency (c), and political party (p) respectively.   Let $\left(0 \le \mu_{B^n}(b), \mu_{B^c}(b) \text{ and } \mu_{B^p}(b) \le 1\right)$ be degrees of accountabilities as expressed toward $A_i^n, A_i^c,$ and A_i^p respectively where $b \in B$ as a cost-benefit value. Let $\mathbb{D}$, the set of members of the social decision-making core with an index set $\mathbb{I}$ and a generic term $i \in \mathbb{I}$, be composed of a number of political parties with an index set, $\left(\mathbb{J} \text{ with } j \in \mathbb{J}\right)$. Let the members of each political party, as a subset of the decision-making core, be represented by $\left(\mathbb{D}_j \subseteq \mathbb{D}, j \in \mathbb{J}\right)$ where $\left(\mathbb{D}_j \cap \mathbb{D}_\lambda = \varnothing \text{ for } j, \lambda \in \mathbb{J} \text{ with } j \ne \lambda\right)$ and $\left(\mathbb{D} = \bigcup_{j \in \mathbb{J}} \mathbb{D}_j\right)$. One party state is defined if $\left(\mathbb{D}_j \subseteq \mathbb{D} \subseteq \mathbb{D}_j, j \in \mathbb{J}\right)$ in which case $(\#\mathbb{J}) = 1$. Similarly, political duopoly is established over the state if $(\#\mathbb{J}) = 2$. We have multiparty democracy if $(\#\mathbb{J}) > 2$. Given $\mathbb{D}$, let there be decision-making committees, $\mathbb{C}$ whose index set is $\mathbb{K}$. Furthermore, let the number of each committee members be $(\#\mathbb{C}_k), k \in \mathbb{K}$ such that $\mathbb{D} = \bigcup_{k \in \mathbb{K}} \mathbb{C}_k$ (note that $\mathbb{D} = \bigcup_{j \in \mathbb{J}} \mathbb{D}_j = \mathbb{D} = \bigcup_{k \in \mathbb{K}} \mathbb{C}_k$). Each member belongs to at least one committee, such that $\left(\left(\mathbb{C}_i \cap \mathbb{C}_j\right) \ne \varnothing, \forall i, j \in \mathbb{K}\right)$. We consider the process of social decision making and accountability in the case of an effective political duopoly where each committee is made up of members from the two parties.

In the social decision-making processes and corresponding accountabilities, we must consider the size of the decision-making core as measured by $(\#\mathbb{D})$, the size of the individual party in the core as measured by $\#\mathbb{D}_j, j \in \mathbb{J}$ and $\sum_{j \in \mathbb{J}} \#\mathbb{D}_j = \#\mathbb{D}$, the number of committees is $\#\mathbb{K}$, the size of each committee is $\#\mathbb{C}_k$, the number of political party members in each committee is $\left(\#\mathbb{C}_{jk}, j \in \mathbb{J} \text{ and } k \in \mathbb{K}\right)$, where all of which constitute the structure of the social decision-making core for the social decision-making process. The social decision-making pattern through individual votes and principle of majority rule is governed by an *accountability process* among different interests as seen by the individual decision-maker relative to the interests of the nation, constituency and political party. The accountability is revealed in subjective and objective cost-benefit or cost-utility imputations at a given responsibility space. Let B be the general net benefit and associated with the nation, n, constituency, c and political party, p be expressed as $\mathsf{B}^n, \mathsf{B}^c$, and B^p respectively. The organizational structure of the social decision-choice process is diagrammatically represented in Figure 5.7 where case of the political duopoly is that $\# \mathbb{J} = 2$ and one party political state is obtained by setting $\# \mathbb{J} = 1$. A multiparty political state is obtained by setting $\# \mathbb{J} > 2$.

In general, we may assume that $\mathsf{B}^n > \left(\mathsf{B}^c \vee \mathsf{B}^p \right)$ where $\left(\vee \Rightarrow \max \right)$, but the individual assessment, given the parameters of choice, may rank them differently in accordance with their utility imputations. For any individual $i \in \mathbb{I}$, we have $u_i \left(\mathsf{B}^n \right) > \left[u_i \left(\mathsf{B}^c \right) \vee u_i \left(\mathsf{B}^p \right) \right]$ or $u_i \left(\mathsf{B}^n \right) < \left[u_i \left(\mathsf{B}^c \right) \vee u_i \left(\mathsf{B}^p \right) \right]$. The ranking of the voting by individual members in the decision-making core is based on the subjective assessment of the individual preference ranking over perceived net benefit for the nation, constituency and party through utility imputations. The preference ordering is such that given B^n, there are three parameters β^n, β^c and β^p that allow an individual to relate national benefits to the benefits of the constituency and party. For an individual $i \in \mathbb{I}$ we have $u_i \left(\mathsf{B}^n \mid \beta^n \right)$, $u_i \left(\mathsf{B}^n \mid \beta^c \right)$ and $u_i \left(\mathsf{B}^n \mid \beta^p \right)$. In other words, the individual assesses the relative national net benefit to the nation, constituency and party. The parameters β^n, β^c and β^p define the environment for individual assessments or preferences over the nation, constituency and party regarding the net benefit to the nation, the relative usefulness of an element in the goal-objective set to the interest of the nation, constituency and party in the democratic collective decision-choice system.

The social decision-voting process is guided by the individual ranking on the basis of $u_i \left(\mathsf{B}^n \mid \beta^n \right)$ as compared to $u_i \left(\mathsf{B}^n \mid \beta^c \right)$ and $u_i \left(\mathsf{B}^n \mid \beta^p \right)$. The parameter β^n captures the environment containing responsibility to the national public, conditions of national interest, welfare, integrity and security. The parameter β^c defines the conditions and environment of the incumbent's re-election, contribution of funds for re-election, personal commitment to constituency and personal gains. The parameter β^p reflects the conditions and environment of party recognition, climbing to levels of seniority, commitment to party ideology, and degree of ascendency on the party's hierarchy amidst a host of personal aspirations and interest. In this discussion, B^n is taken to be a reasonably objective measure, while the composite values β^n, β^c and β^p are subjective elements that affect the valuation of B^n as well as constitute the knowledge bases of vote-casting in terms of accountability conditions on behalf of the members of the decision-making core given the social responsibility environment.

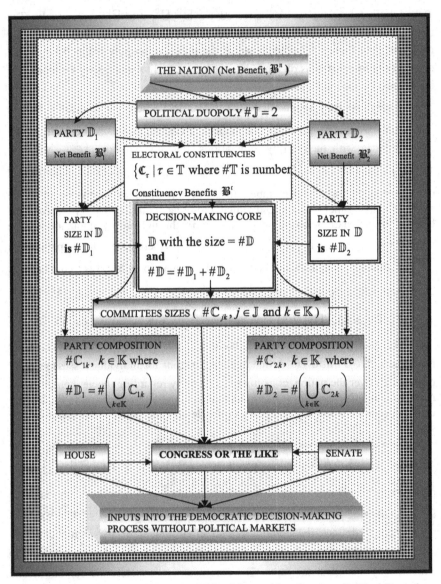

Fig. 5.7 The Internal structure of the Social Decision-making Core in a Political Duopoly

5.2.2.1 The Social Decision-Choice Process, Cost-Benefit Rationality and Social Accountability

Let $V_i(n)$ be a voting function for individual $i \in \mathbb{I}$ and degrees of accountability expressed in terms of fuzzy preferences as $\mu_{A^n}(b)$, $\mu_{A^c}(b)$ and $\mu_{A^p}(b)$, $b \in A$ which is individually assessed and defined by the way of net benefit condition associated with three states of the nation, the constituency and the political party for every important bill related to national interest.

$$V_i(n) \begin{cases} = 1 \text{ if } u_i\left(B^n, \beta^n\right) > \left[\left(u_i\left(B^n, \beta^n\right)\right) \vee \left(u_i\left(B^n, \beta^n\right)\right)\right] \Rightarrow \mu_{A^n}(b) > \left[\left(\mu_{A^c}(b)\right) \vee \left(\mu_{A^p}(b)\right)\right] & i \in \mathbb{I} \\ = 0, \text{ otherwise} \end{cases}$$

(5.1)

We may use eqn. (5.1) to define total accountability N to the national interest and the supporting social goal objective set as reflected in the voting space as: $N = \sum_{i \in \mathbb{I}} V_i(n)$ with a complement of $N' = \#\mathbb{I} - \sum_{i \in \mathbb{I}} V_i(n)$. The degree or an index of political accountability to the national interest α of the social decision-making core is then specified as $\alpha = \left(N/\#\mathbb{I}\right) \in [0,1]$. The political system's sensitivity to citizen's sovereignty and accountability may be specified as:

$$\left(N/\#\mathbb{I}\right) \begin{cases} > \left(N'/\#\mathbb{I}\right) \Rightarrow \text{Politically accountable to national interest} \\ = \left(N'/\#\mathbb{I}\right) \Rightarrow \text{Politically neutral} \\ < \left(N'/\#\mathbb{I}\right) \Rightarrow \text{Politically accountable to either Conastituency or Party or both} \end{cases}$$

(5.2)

The individual's assessed net cost-benefit is more or less subjective from the perception of relative characteristics of national interest, constituency interest and party interest. From the pattern of voting, we can examine the question of incumbency and party affinity for those that fall into the voting pattern with $N' = \left(\#\mathbb{I} - \sum_{i \in \mathbb{I}} V_i(n)\right)$ in terms of whether given $V_i(n) = 0 \Rightarrow \mu_{A^c}(b) \gtreqless \mu_{A^p}(b)$ where $(>) \Rightarrow$ incumbency affinity, $(<) \Rightarrow$ party affinity and $(=) \Rightarrow$ indifference. These affinities are driven by a *set of motivation elements* M with a generic variable $m \in M$ that directs decision-choice actions over the social decision-choice space. The underlying assumption here is that decision-choice agents are subjectively purposeful and this purposefulness can be influenced by altering the cost-benefit (or cost-utility) balances that affect personal interest.

The question that arises may be stated as: what are the motivational elements in M that are associated with the defined affinities to national interest, constituency interest and party interest? The answer must be found in the elements in the incentive structure, I with a generic element $\eta \in \mathsf{I}$ that is associated with the political decision-choice process. The elements of the incentive structure must appeal to the individual personal net cost-benefit configuration. The incentive structure I may be partitioned into monetary I_m and non-monetary I_n incentive sub-structures, such that $\mathsf{I}_m \cap \mathsf{I}_n = \varnothing$ and $\mathsf{I}_m \cup \mathsf{I}_n = \mathsf{I}$. For democratic decisions without a political market, it is reasonable to assume away the monetary incentive and work with a set of non-monetary incentives I_n. The non-monetary incentive structure may be defined to include moral incentives, social coercive incentives, social prestige and social admiration. All elements in the non-monetary incentive structure may apply to the three categories of the socio-political interest. The provision of more responsibility and social prestige that is associated may constitute an important incentive to decide in favor for national interest and hence vote for legal, social and economic programs and projects that support national interest and a set of supporting programs and its required best distribution, even if they go against the interests of the constituency, political party of one's belonging and one's incumbency interest. This incentive structure may be complicated if responsibility and committee assignments are done on seniority basis in which case conditions of incumbency and the interests of the constituency become overriding incentives to decide.

Let $\mathsf{R}_i(\eta)$ be an incentive response function of individual $i \in \mathbb{I}$ to an incentive $\eta \in \mathbb{I}_n$ then incentive efficiency may be specified if $\mathbb{I} = \mathbb{I}_n \cup \mathbb{I}'_n$ where the index set of those voting against national interest but in favor of either interest of the constituency or the political party is $\mathbb{I}'_n$. The response function is such that:

$$\mathsf{R}_i(\eta) \begin{cases} = 1, \text{ for possitive incentive response} \\ = 0, \text{ for non-positive incentive response} \end{cases}, \eta \in \mathsf{I}_n, i \in \mathbb{I}'_n \qquad (5.3)$$

Similarly,

$$\mathsf{V}_i(\mathsf{n}) \begin{cases} = 1, \text{ if } \mathsf{R}_i(\eta) = 1 \ i \in \mathbb{I}'_n \\ = 0, \text{ if } \mathsf{R}_i(\eta) = 0 \ i \in \mathbb{I}'_n \end{cases}, \eta \in \mathsf{I}_n \qquad (5.4)$$

The incentive response function reveals to us those incentive elements that help to shape accountability to the national interest given the responsibility space. The accountability to national interest must not be interpreted as conditions of patriotism even though there are tendencies to project and associate such a conclusion. The aggregate accountability index will increase to the extent to which the response function is positive and effective. The incentive structure works through moral persuasion and commitment to national progress and a belief in the established national interest. The political accountability and the incentive structure are carried to the national budgetary decisions in support of the goal-objective set to actualize or maintain the national interest. We shall now turn our attention to the possible relationship between accountability and the budgetary process.

5.3 Political Accountability and the Budgetary Decision-Choice Activities in Non-political Markets

Here, the accountability may be related to governance by objectives or governance by personal interest. Let us keep in mind that the existence of institutions of government creates principal-agent duality and the corresponding principal-agent problem that is often magnified by information asymmetry in relation to social budgetary decisions to implement projects in support of the elements in the social goal-objective set. We must also keep in mind that budgetary decisions may be influenced by the ideological positions on the social visions and national interests that one holds which will come to influence accountability. Now let us examine the political accountability and the budgetary decisions given a non-political market in democratic social formation in relation to a democratic collective decision-choice system. It may be pointed out that there is no political accountability if there is no socio-political responsibility. In other words, the accountability space must be mapped onto the responsibility space. The political accountability in the rims of the budgetary process requires assignments of responsibility of the budget designs and budget implementation to the members of the decision-making core. The budgetary decisions involve determining the size of the overall budget, the projects to be funded or to be discontinued, the budget allocation decision over projects and the distribution of projects over constituencies in support of the elements of the social goal-objective set.

   The political accountability process will be affected by the manner in which the overall budget is constructed. The overall budget may be constructed in the following two ways: 1) from the overall budget size to a disaggregation into project-constituency distributions or 2) from the project-constituency distributions to an aggregation into the overall budget size. The former is from general to specific where the macro-budgetary decisions precede the micro-budgetary decisions; and the latter is from specific to general where the micro-budgetary decisions precede the macro-budgetary decisions. In the general-to-particular approach, the macro-budgetary decision is an enveloping of the micro-budgetary decisions. In the particular-to-general approach, the micro-budgetary decisions are approximations to the macro-budgetary decision. The structure of accountability relative to the social budgetary decisions may be presented in a form of cognitive geometry as in Figure 5.8.

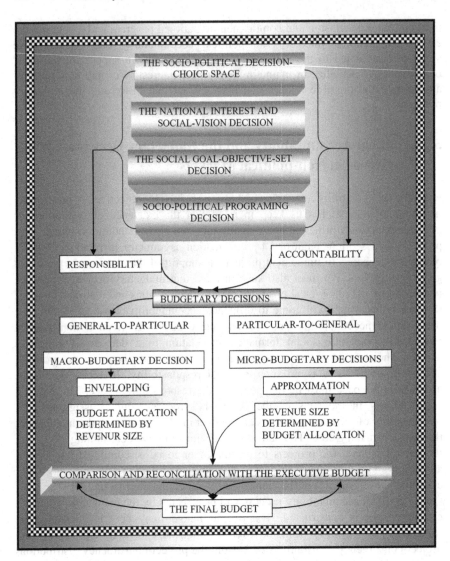

Fig. 5.8 A Cognitive Geometry of the Logical Flow of the Budgetary-Decision Process

These general-to-particular and particular-to-general budgetary operational systems are creatively linked to a revenue-generating system. If the decision-making core adopts the general-to-particular approach then the budget size is fixed by our revenue-generating system for a balanced budget and the revenue pie is to be slashed among budget categories. In this respect, the revenue-generating decision precedes the budget-allocation decision where the size of the revenue becomes a binding constraint on the budgetary decisions that will affect the micro-budgetary decisions. The micro-budgetary decision is simply an allocation efficiency problem the solution of which must be guided by an efficiency criterion. On the other hand, if the decision-making core adopts the particular-to-general approach, then the revenue size is determined by the budgetary allocation system where the sources of revenue are to be identified in the legislative process for a balanced budget. Here, the aggregation of the micro-budgetary decisions leads to the required overall budget. In an integrated and interdependent system, the two approaches may be considered as constraints on each other. The manner in which needs and budget information are developed will emerge out of responsibility which will then come to influence the accountability of the members of the decision-making core. In all socio-political systems, there are budget categories and programs and host agencies. Programs are housed in budget categories as specified and agreed upon by the members of the decision-making core after the budget's passage into an expenditure bill.

5.3.1 Modeling Accountability and the Budgetary Decisions

Let us now present a structure of the budgetary decisions that will allow us to examine the accountability question. Let $\mathbf{L}$ with generic element $\ell \in \mathbf{L}$ be the index set of numbers of budget categories $\mathscr{B}$. Furthermore, let $\mathbf{K}$ with generic element $k \in \mathbf{K}$ be the index set of programs $\mathscr{P}$ from the goal objective set. To examine responsibility and accountability, we have before us the following decision making sets to work with.

A. The Electoral Constituency Set

$$C = \left\{ C_\tau \mid \tau \in \mathbb{T} \text{ where } \mathbb{T} \text{ is the index set of constituencies and } C = \text{ The Nation } \right\}.$$

(5.5)

Politically, a nation is a set of constituencies. As an example, we may consider the states within the United States of America as constituting a set of constituencies C_τ which runs over the states that themselves are divided into districts. Thus, C_τ is a set of Districts where $C = \bigcup_{\tau \in \mathbb{T}} C_\tau$ is the nation and $\bigcap_{\tau \in \mathbb{T}} C_\tau = \varnothing$.

B. The Party Composition Set

$$\mathbb{D} = \left\{ \begin{array}{l} \mathbb{D}_j \mid j \in \mathbb{J}, \text{ index set of political parties and } \mathbb{D} = \bigcup_{j \in \mathbb{J}} \mathbb{D}_j \text{ and } \bigcap_{j \in \mathbb{J}} \mathbb{D}_j = \varnothing \\ \text{where in political duopoly } \#\mathbb{J} = 2 \end{array} \right\}$$

(5.6)

In this case, using the example of the United States of America, $\mathbb{D}$ constitutes the congress which may be divided into the House of Representatives $\widehat{\mathbb{D}}$ and the Senate $\breve{\mathbb{D}}$ where $\mathbb{D} = \widehat{\mathbb{D}} \cup \breve{\mathbb{D}}$ and $\widehat{\mathbb{D}} \cap \breve{\mathbb{D}} = \varnothing$. In other words, the responsibility space is uniquely partitioned.

The Set of the House of Representatives

$$\widehat{\mathbb{D}} = \left\{ \widehat{\mathbb{D}}_j \mid j \in \widehat{\mathbb{J}}, \text{ index set of political parties and } \widehat{\mathbb{D}} = \bigcup_{j \in \widehat{\mathbb{J}}} \widehat{\mathbb{D}}_j \text{ and } \bigcap_{j \in \widehat{\mathbb{J}}} \widehat{\mathbb{D}}_j = \varnothing \right\}$$

(5.7)

The Set of the Senate

$$\breve{\mathbb{D}} = \left\{ \breve{\mathbb{D}}_j \mid j \in \breve{\mathbb{J}}, \text{ index set of political parties and } \breve{\mathbb{D}} = \bigcup_{j \in \breve{\mathbb{J}}} \breve{\mathbb{D}}_j \text{ and } \bigcap_{j \in \breve{\mathbb{J}}} \breve{\mathbb{D}}_j = \varnothing \right\}$$

(5.8)

C. The Committee Formation Set for Responsibilities and Decision-Making

$$\widehat{\mathbb{D}} = \left\{ \widehat{\mathbb{C}}_k \mid k \in \mathbb{K}, \text{ index set of committees and } \widehat{\mathbb{D}} = \bigcup_{k \in \mathbb{K}} \widehat{\mathbb{C}}_k \text{ and } \bigcap_{k \in \mathbb{K}} \widehat{\mathbb{C}}_k \neq \varnothing \right\}$$

(5.9a)

$$\breve{\mathbb{D}} = \left\{ \breve{\mathbb{C}}_k \mid k \in \mathbb{K}, \text{ index set of committees and } \breve{\mathbb{D}} = \bigcup_{k \in \mathbb{K}} \breve{\mathbb{C}}_k \text{ and } \bigcap_{k \in \mathbb{K}} \widehat{\mathbb{C}}_k \neq \varnothing \right\}$$

(5.9b)

For administrative efficiency and simplification of the tasks of decision-making and implementation of elements in the goal-objective set, the decision-making core is divided into committees $\left(\mathbb{C}_k, \ k \in \mathbb{K} \right)$ for the established areas of the decision-choice process. Each committee corresponds to a defined national decision-making area where each member is assigned to a committee. For example, when one takes the United State of America, there are twenty committees and about 68 sub-committees for the Senate. These may change depending on social dynamics, national interests, social needs and social-problem solving requirements for social management at each social-decision period. The term $\mathbb{C}_k, \ k \in \mathbb{K}$ is a committee with k representing the Committee on Agriculture, Nutrition and Forestry. The subscribe k runs over all currently existing committees such as committees on Budget, Foreign Relations, Armed

Services and others. In other words, $\mathbb{C}_k$ is a set of individuals working on decision category $k \in \mathbb{K}$ as organized by the decision-making core to execute the organic decision-choice process in the political economy. The use of the social structure of USA is just an illustrative example provided for the understanding of the general theory that is being advanced on the political economy.

C.1 The Committee Assignment Set

$$\hat{\mathbb{C}}_k = \left\{ \hat{\mathbb{C}}_{jk} \mid j \in \mathbb{J} \text{ with fixed } k \in \mathbb{K} \text{ where } \mathbb{J} \text{ is an index set of parties, } \hat{\mathbb{C}}_k = \bigcup_{i \in \mathbb{J}} \hat{\mathbb{C}}_{jk} \text{ and } \bigcap_{i \in \mathbb{J}} \hat{\mathbb{C}}_{jk} = \varnothing \right\}$$
(5.10a)

$$\breve{\mathbb{C}}_k = \left\{ \breve{\mathbb{C}}_{jk} \mid j \in \mathbb{J} \text{ with fixed } k \in \mathbb{K} \text{ where } \mathbb{J} \text{ is an index set of parties, } \breve{\mathbb{C}}_k = \bigcup_{i \in \mathbb{J}} \breve{\mathbb{C}}_{jk} \text{ and } \bigcap_{i \in \mathbb{J}} \breve{\mathbb{C}}_{jk} = \varnothing \right\}$$
(5.10b)

The members of the decision-making core are then assigned to each committee $\mathbb{C}_k$, $k \in \mathbb{K}$ given the defined rules of assignment. For example, in the political duopoly of the United States of America, in the case of the Senate Committee on Agriculture, Nutrition and Forestry, $\mathbb{C}_{kd}$ is the set of Democratic members and $\mathbb{C}_{kr}$ is the set of Republican members in the committee $\mathbb{C}_k$ where $\mathbb{C}_k = \mathbb{C}_{kd} \cup \mathbb{C}_{kr}$ and $\mathbb{C}_{kd} \cap \mathbb{C}_{kr} = \varnothing$. The same will apply to the House Committees given its rules of committee formation and assignments.

C.2 The Budget Category Set

$$\mathscr{B} = \left\{ \mathscr{B}_\ell \mid \ell \in \mathsf{L}, \text{ the index set of budget categories where } \mathscr{B} = \bigcup_{\ell \in \mathsf{L}} \mathscr{B}_\ell \text{ and} \bigcap_{\ell \in \mathsf{L}} \mathscr{B}_\ell = \varnothing \right\}$$
(5.11)

An illustrative example of the budget category set is that of the United States of America with $\# \mathsf{L} = 20$ as current budget categories with National Defense (050), International Affairs (150),..., Community and Regional Development (450),...., General Government (800) and others budgetary decision categories.

C.3 The Program Set and the Budget Home

$$\mathscr{P} = \left\{ \mathscr{P}_k \mid k \in \mathsf{K} \text{ and } \mathscr{P}_k \in \mathscr{B}_{\ell k}, \ell \in \mathsf{L} \right\}$$
(5.12)

$\mathscr{P}_k$ is the kth Program located in the budge home $\mathscr{B}_{\ell k}$, $\ell \in L$. For example, in the case of the Unite State of America, $\mathscr{P}_k$ is a set composed of activities of National Science Foundation, National Aeronautics and Space Administration and general science programs at Department of energy. $\mathscr{B}_{\ell k}$ $\ell \in L$ is the budget category of General Science, Space and Technology.

C.4 The Budget Responsibility Assignment Set

$$\widehat{\mathscr{R}} = \left\{ \widehat{\mathscr{R}}_{\ell i_{j\tau}} \mid i \in \widehat{\mathbb{I}}, \, \ell \in L, \, j \in \widehat{\mathbb{J}} \text{ and } i \in \mathcal{C}_{\tau}, \, \tau \in \mathbb{T}, \quad \text{where } \widehat{\mathscr{R}} = \bigcup_{\ell \in L} \widehat{\mathscr{R}}_{\ell j\tau} \text{ and } \widehat{\mathscr{R}}_{\ell i_{j\tau}} \subseteq \widehat{\mathbb{D}} \right\}$$
(5.13)

where $\widehat{\mathbb{I}}$ is an index set of the House of Representatives, and $\widehat{\mathbb{J}}$ is an index set of party affiliations of the House members.

$$\breve{\mathscr{R}} = \left\{ \breve{\mathscr{R}}_{\ell i_{j\tau}} \mid i \in \breve{\mathbb{I}}, \, \ell \in L, \, j \in \breve{\mathbb{J}} \text{ and } i \in \mathcal{C}_{\tau}, \, \tau \in \mathbb{T}, \quad \text{where } \breve{\mathscr{R}} = \bigcup_{\ell \in L} \breve{\mathscr{R}}_{\ell j\tau} \text{ and } \breve{\mathscr{R}}_{\ell i_{j\tau}} \subseteq \breve{\mathbb{D}} \right\}$$
(5.14)

where $\breve{\mathbb{I}}$ is an index set of the Senate, and $\breve{\mathbb{J}}$ is an index set of party affiliations of the Senate members.

The set $\widehat{\mathscr{R}}$ represents the House, and the set $\breve{\mathscr{R}}$ represents the Senate. The term $\mathscr{R}_{\ell i_{j\tau}}$ is a set of members $i \in \mathbb{I}$ from the social decision-making core and constituency $\left(i \in \mathcal{C}_{\tau}, \tau \in \mathbb{T} \right)$, with the index set of constituencies with party affiliation $j \in \mathbb{J}$ assigned to be responsible in developing budget information on budget category $\mathscr{B}_{\ell}$ with $\ell \in L$. The information developed by the members of the budget Responsibility Assignment Set is then transmitted to the budget committees of the social decision making core (for example the House and Senate in the case of the United States of America) for the final write up and vote taking by all members of the social decision-making core under the established rules of vote counting, where disagreements may be reconciled under established rules of reconciliation. The Executive Branch develops its own budget in relation to the national interest, social vision, the goal-objective set and required programs as conceived in the executive branch to accomplish both the domestic and international interests as have been discussed. The public budgetary decisions and their implementations can affect the character of the nation, its general welfare and the happiness of the people.

Once again, the decision-making pattern through individual votes and principle of majority rule is governed by an accountability process among interests as seen by the individual decision-maker relative to the nation, constituency and political party as revealed in subjective cost-utility and objective cost-benefit imputations of the total budget. The decision to approve the national budget becomes entangled in national, party and constituency politics that can create undesirable outcomes and unintended consequences for the public which is being represented as the principal and by the decision-making core as the agent. A question, therefore, may be asked: what does an individual vote express? Suppose an individual votes against (for) a national budget bill, does that mean one is not accountable (accountable) to the national aspirations as expressed in the budget categories which contain the program sets in relation to the goal-objective set in support of the national interest? Let $B(\mathscr{B})$ define the net social benefit over cost associated with the budget $\mathscr{B}$ and let the net benefits associated with the nation, n, constituency c and political party p be expressed as $B^n(\mathscr{B})$, $B^c(\mathscr{B})$, and $B^p(\mathscr{B})$ where the budget is to support the managing of the goal-objective set intended for accomplishing the national interests and social vision . The vote of an individual in the decision-making core is cast in relation to comparative assessments of the relative preference strengths of net cost-benefit relations of the nation, constituency and party affiliation. At this point, we must distinguish among the cost-benefit imputations, cost-utility imputations and cost-risk imputations in the subjective decision processes at both the macro and micro-decision levels.

5.3.2 Political Accountability, the Budgetary Decisions and the Voting Process with Fuzzy Preference Relations

Let $V_i(n)$ be a voting function for individual $i \in \mathbb{I}$ and degrees of accountability expressed in terms of fuzzy preference as $\mu_{A^n}(b)$, $\mu_{A^c}(b)$ and $\mu_{A^p}(b)$ which is individually assessed and defined on the basis of net benefit conditions associated with the three states of the nation, constituency and political party for the budget composition. The voting function may be specified as in eqn.(5.15) and (5.16)

$$V_i(n) \begin{cases} = 1 \text{ if } u_i\left(B^n,\beta^n\right) > \left[u_i\left(B^n,\beta^n\right) \vee u_i\left(B^n,\beta^n\right) \right] \Rightarrow \mu_{A^n}(b) > \left[\mu_{A^c}(b) \vee \mu_{A^p}(b) \right] \ i \in \mathbb{I} \\ = 0 \text{, otherwise} \end{cases}$$

$$(5.15)$$

$$V_i(n) \begin{cases} = 1 \text{ if } B_i^n(\mathscr{B}) > \left(B_i^c \text{ or } B_i^p \right) \Rightarrow \mu_{A^n}(b) > \left(\mu_{A^c}(b), \mu_{A^p}(b) \right) \ i \in \mathbb{I} \\ = 0 \text{ otherwise} \end{cases}$$

$$(5.16)$$

We may use eqns. (5.15) and (5.16) to specify the total accountability N to the national interest and goal objective set as reflected in voting as: $N = \sum_{i \in \mathbb{I}} V_i(n)$ as the number of those who vote for national interest with a complement of $N' = \#\mathbb{I} - \sum_{i \in \mathbb{I}} V_i(n)$ for those who vote on the bases of either party affinity or constituency affinity. The degree or an index of political accountability α of the decision-making core may then be specified as $\alpha = \left(\frac{N}{\#\mathbb{I}} \right)$. The political system's sensitivity to citizen's sovereignty and accountability may be specified, given that a yes vote is good for the nation viewed in terms of social cost-benefit balances as:

$$\left(\frac{N}{\#\mathbb{I}} \right) \begin{cases} > \left(\frac{N'}{\#\mathbb{I}} \right) \Rightarrow \text{Politically accountable to national interest} \\ = \left(\frac{N'}{\#\mathbb{I}} \right) \Rightarrow \text{Politically neutral} \\ < \left(\frac{N'}{\#\mathbb{I}} \right) \Rightarrow \text{Politically accountable to either Constituency or Party} \end{cases}$$

$$(5.17)$$

The central assumption here is that the net cost-benefit configuration of the national budget as reconciled is approximately optimal and greater than either the net cost-benefit configurations of those of the constituency and party affiliation as may be technically assessed, given the goal-objective set in support of the national interest. There is a further assumption that the members in the decision-making core are in agreement on the national interest configuration. Any disagreement on the national interest configuration from some of the members of the decision-making core may result in intense budgetary disagreement. The decision-voting comparisons may be done with individual cost-benefit or cost-utility configurations relative to their assessment of the national budget. The individuals are said to be responsible and accountable to the nation if their voting decisions correspond to negative net cost-benefit configurations of the national budget even if their imputations are incorrect.

How do we view the relationships between voting and accountability to the nation? In other words, how do we interpret voting by individual members of the decision-making core in relation to positional accountability to nation, party, constituency and self? Is commitment to party platform leading to a negative vote on a bill considered as an accountability or non-accountability to the nation? When shall we say that the individual in the decision-making core or the core in general is not accountable and to what? There are many such disturbing questions but the answers to these questions must be examined around both non-budgetary and budgetary matters of the public. The complexity surrounding the analysis of *social responsibility* and *political accountab*ility of the members of the decision-making core requires methods of *synergetics* and complexity that may be amplified by our understanding of the methods in *energetics*. The answers to these questions will depend on how the decision-making core is conceived in theory and practice. If it is viewed as a national decision-making body whose preferences are shaped by the national cost-benefit creation, then the members are elected to represent the majority view of the nation on national interests and the goal-objective set needed to support the best cost-benefit bundle of the nation.

The nation is the general public, which constitutes the principal, to which the social decision-making core that constitutes the agent, is accountable. Thus an elected member of the decision making-core is a mouthpiece of his or her constituency for the democratic decision-making in the sense of presenting the essentials of the collective view of his or her constituency regarding the interests of the nation. In this process, the political accountability is to the nation and must be exercised in the best understanding of what the national interest is or should be. Social decisions must be made to optimize the net social cost-benefit bundle for the whole society. The construct of the goal-objective set in support of the national interest must be guided by the principle of constituency fairness and in a manner that enhances the optimal welfare of the nation.

The set of projects that flow from the implementation of the elements in the goal-objective set must ideally be distributed over the constituencies in term of need and their impacts on the welfare of the nation and its people where there are implied *fairness criteria* that will help to do away with uneven development. It is here that constituency accountability and national accountability tend to crash over budgetary decisions needed to produce budget distribution to finance the mandates which are implied in the program set. It is also, here, that who gets what, when and where becomes a challenge of the application of fairness. A reasonable way to examine accountability of the decision-making core to the public is to design two voting systems. One voting system captures the preferences of the public, and the other captures the preferences of the decision-making core on a given social decision-choice item. The directions of individual decisions in the decision-making core become related to the incentive structure in support of public decision-making. From equation (5.15), the α may also be constructed from the voting preferences of the general voting public.

$$V_\ell(n) \begin{cases} = 1 \text{ if } u_\ell\left(B^n, \beta^n\right) > \left[u_\ell\left(B^n, \beta^n\right) \vee u_\ell\left(B^n, \beta^n\right)\right] \Rightarrow \mu_{A_\ell^n}(b) > \left[\mu_{A_\ell^c}(b) \vee \mu_{A_\ell^p}(b)\right] \ell \in \mathbb{L} \\ = 0 \text{ , otherwise} \end{cases}$$

$$(5.18)$$

The symbol $\mathbb{L}$ is an index set of the voting public, $Q = \sum_{\ell \in \mathbb{L}} V_\ell(n)$ is the proportion of the public that vote in favor, and $Q' = \#\mathbb{L} - \sum_{\ell \in \mathbb{L}} V_\ell(n)$ is those who vote against relative to the social cost-benefit configuration. We may thus structure the system's accountability as:

$$\left\{\begin{aligned}\left(\frac{N}{\#\,\mathbb{I}}\right)&\geq\left(\frac{Q}{\#\,\mathbb{L}}\right)>\left(\frac{Q'}{\#\,\mathbb{L}}\right)\\\left(\frac{N'}{\#\,\mathbb{I}}\right)&\geq\left(\frac{Q'}{\#\,\mathbb{L}}\right)>\left(\frac{Q}{\#\,\mathbb{L}}\right)\end{aligned}\right\}\Rightarrow \text{Politically accountable to the national interest and the public}$$

$$(5.19)$$

Similarly,

$$\left\{\begin{aligned}\left(\frac{N}{\#\,\mathbb{I}}\right)&<\left(\frac{Q}{\#\,\mathbb{L}}\right)>\left(\frac{Q'}{\#\,\mathbb{L}}\right)\\\left(\frac{N'}{\#\,\mathbb{I}}\right)&<\left(\frac{Q'}{\#\,\mathbb{L}}\right)>\left(\frac{Q}{\#\,\mathbb{L}}\right)\end{aligned}\right\}\Rightarrow \text{Politically non-accountable to the national interest and the public}$$

$$(5.20)$$

Equations (5.19) and (5.20) are simple tests of accountability of the same information structure available to both the public as the principal and the social decision-making core as the agent. In other words, there is no information asymmetry in the democratic decision-making process between the principal and the agent. The accountability is defined and measured in terms of majority rule. The members of the public, just as the members of the decision-making core, have information-processing capacity to assess the cost-benefit balances regarding the decision-choice element under vote. In this way, the decision-making core is accountable if the members' decision action is in line with that of the public majority. This test can be translated into cost-benefit calculations and rationality. The principle of majority rule in the democratic collective decision-making system has an implicit assumption that each voting member has the same information about the net cost-benefit values, net cost-utility values or net cost-risk values. The acceptance of the principle simply suggests that the aggregate net cost-benefit value of the majority is greater than the aggregate net cost-benefit value of the minority. This principle in managing the social decision-choice system is at risk if there are political and legal markets that encourage buying and selling of influence in order to foster rent-creation, rent-preservation and rent harvesting. The risk to democracy is further increased if the members of the voting public are forced to operate with a *deceptive information structure* that is composed of disinformation, misinformation and information classification on the basis of national security to disadvantage the voting public to operate in the sections of the information sub-space. It is, here, that the concept of political transparency acquires content and meaning. It is also, here, that the phenomena of social responsibility and political accountability meet the phenomenon of rent-seeking and where rent-seeking generates dynamics of change.

Chapter 6
Political Markets and Penumbral Regions of Rent-Seeking Activities

Let us examine the relational structures of the social collective decision-choice activities and political markets. This will be followed up with rent-seeking activities as part of the incentive structure of rent-seekers travelling along the penumbral regions of decision-choice actions and choice of strategy for institutional transformations.

6.1 Rent-Seeking as a Driving Force for Changes in Political and Legal Institutions

An organized society, under constitutional democracy and capitalism, may be viewed in terms of a capital-owning class who controls the ownership of capital and non-capital-owning class who owns of labor services. The capital-owning class may be seen in terms of a rent-seeking class and a non-rent seeking class for the capitalists. This is particularly so in the Schumpeterian political economy of bureaucratic capitalism with deep-seated plutocracy. The nature of rent-seeking, the structure of rent-creation, the manner of rent-preservation, the degree of protection accorded to the rent-seeking class and the distribution of rent-harvesting possibilities depend on the legal restrictions, the constitution and its effects on socio-economic liberties. The socio-economic liberties define the dynamics of the individual-community duality as it affects the distribution of opportunities, income and rights in the social decision-making space of the political economy. Let us keep in mind that the legal restrictions as constraints on decision-choice actions are defined in the legal structure. The legal structure specifies the extensive and intensive regulatory regimes to create fairness and unfairness in the distributions of justice, income, opportunities and many others elements that contain cost-benefit characteristics. Rent-seeking is intense in societies whose constitution emphasizes individual interests and freedoms over collective interests and freedoms. Rent-seeking is less intense and reduced to a reasonable minimum in societies that emphasize collective interests and freedoms over those of the individual. The pursuit of individual interests drives the rent-creation and rent-preservation process. The pursuit of individual freedoms without regard to that of the collective, given rent creation and preservation, drives the rent-harvesting process to the extent to which the regulation intensity is practiced, given the regulation widening.

The driving force of the rent-seeking process is derived from the principle of individual interests and freedoms. The rent-seeking phenomenon is the product of a social organism, under law, rules and regulations, where the social organism seeks to restrain the practice of the principle of individualism relative to personal interests and freedoms. After the birth of the rent-seeking process, it acquires its own identity and living essence whose development changes the institutions of the social organism, while its own development is affected by the changing institutions of the social organism. The dynamics may be seen from the conflicts in the interest behavior of the individual-community duality, where the principle of individualism seeks to assert its dominance over the principle of collectivism in relation to the distribution of opportunities, resources and income that are broadly defined in the political economy. The changes in the institutions of the social organism are both qualitative and quantitative where the quality may be seen in terms of the characters of the individual institutions, and the quantity may be seen in terms of the number of net additions to the existing institutions. Both the quality and quantity of the institutions combine to shape the nature and character of the overall institutional configuration and the policy transmission mechanism of the social organism.

The process of rent-seeking, given the partition of society into a rent-seeking class and a non-rent-seeking class, is put in motion by the decision-choice conflicts between the rent-seeking class as a segment of the principal and the social decision-making core as the socio-political agent. Let us also keep in mind that the rent-seeking class is also divided into an innovation-investment rent-seeking class and a non-innovation-investment rent-seeking class. Analytically, we may speak of rent production, rent consumption and rent-seeking markets in the rent-seeking process. The rent production is in the political market, the rent-protection is in the legal market and the rent consumption in terms of harvesting is in the economic market. In this way the three markets are relationally connected to determine the nature of income-wealth distribution in the social set-up. The changes in events in the political market alter the events in the legal market and manifest themselves in the economic market as income distribution that further intensifies the demarcation of the social class divisions as well as increases the social will to create new institutions and destroy outdated social institutions. In other words, the profit-seeking process in bureaucratic capitalism is complemented by the rent-seeking process in the political and legal structures whose behaviors are revealed in the economic market. The social institutions are the vehicles through which policies are transmitted to affect the general income-wealth distribution mechanism in the political economy. As discussed, the creation and maintenance of the social institutions are dependent on the distribution of the decision-making power. The same social institutions determine the power distribution in the social decision-space. As such, in the last analysis, the income-wealth distribution is determined by the decision-power relations in the political economy. In this way, poverty is institutionally generated and hence, its solution is derivable from the institutional structure that generates it. When the global data are examined, where data on poverty and corresponding institutional structures on different social set-ups are obtained, we shall find that poverty is related to the nature of institutional configuration, governance and power distribution, given appropriate measures.

The rent-seeking class organizes private institutions to open the doors of influence-peddling in the political space, rent-creation and rent-protection in the legal space, and rent-harvesting in the economic space. The social decision-making core (the occupiers of government) organizes public institutions to close the doors of influence- peddling in the political space, rent-creation and rent-protection in the legal space, and rent-harvesting in the economic space. The extent to which the government is successful depends on the integrity of the governance as a neutral body. The more the government creates public institutions to close the doors, the more new private institutions are created by the rent-seeking class to overcome the institutional constraints of the government, and the more corrupt the political and legal markets become. The more new private institutions are created by the rent-seeking class, the more the government tightens the qualitative and quantitative institutional constraints on the activities of the rent-seekers through public institutions and different regulatory regimes. These interactive dynamics in the public-private interest relations defining institutional changes in the political economy find analytical expressions in the principles of opposites, social polarity, logical duality, conflicts, continuum, social synergetics, unity and systemic uncertainty. It is the presence of the continuum that connects the two opposites in the duality to define a unity. It is the conflicts of the opposites that generate energy and provide forces of transformation in the quantity-quality social disposition. It is here that the classical and fuzzy paradigms must be examined in terms of their overall usefulness in understanding the political economy as a dynamic system that is self-exiting, self-correcting self-moving and self-integrating in a synergetic manner through the collective decision-choice system. One may look at the supply-demand analytical structure as a simplified dualistic cost-benefit understanding of the political economy with a given institutional configuration in static and dynamic cases.

The *classical paradigm* can tell us whether a change has occurred or not, but it is not helpful in dealing with the continuum of qualitative transformation since its laws of thought operate on the principles of non-contradiction and the excluded middle where whatever has happened is true or false but not both. We can speak of institutional change but not the process of the institutional change that alters the quality of the social organism. Every institutional configuration presents a qualitative equilibrium state under tension and plenum of social forces. Then understanding requires a logic that can handle phenomena of quality, quantitative and subjectivity. The use of the classical paradigm, in the construction of current economic theories and analysis, is limited in this respect. The *fuzzy paradigm* tells us whether a change is occurring between the opposites, in what degree and also in what degree can we classify the change, as well as whether a complete change has occurred or not. It is also helpful for us to be able to examine and analyze the decision-choice process and the dynamics of the socio-political game events in the continuum, since its laws of thought operate on the principles of contradiction and the non-excluded middle in the sense that truth and false reside in an inseparable unity. In the fuzzy logical framework every truth has false support and every false has truth support. In other words, within the definition of the interest and freedom in relation to the private-public duality, rent-seeking becomes the driving force of institutional dynamics of the political economy. These institutional dynamics lead to a sequence of *voting games* which is played by the political parties in terms of responsibility, accountability, allegiance and good governance. The strength of introducing the fuzzy paradigm with its logic and mathematic in the study of social phenomena is the analytical flexibility that it brings to handle conditions of complexity, synergetics, energetics quality and subjectivity given the objective and quantitative characteristics of information structures.

6.2 A Sequence of Political Decision-Choice Games in a Political Duopoly

The sequence of the political decision-choice games in relation to institutional creation, policy manifestation, transmission mechanism, revenue generation and distribution, rule-making and other political activities is simply the voting game of the members of the social decision-making core after its constitution. The responsibility and accountability of the general public are the participation to create the social decision-making core and the participation in funding the business of the core. The first of the sequence of the decision-choice games is the determination of the size of the government where the corresponding game is the *small-big government game*.

6.2.1 The Small-Big Government Game

The structure of this small-big government game is discussed in Chapter 3 and is related to the Marxian and Schumpeter types of socialism as shown in Figure 3.1.1, while the characteristics of the Schumpeterian and Marxian political economies are discussed in [R10.15]. Similarly, the diagram with points $A, \left(\rho_\Pi^r = \left(\pi_\Pi^r, \beta_\Pi^r \mid \delta_\Pi^I, \delta_P^I \right) \right)$ and $B, \left(\rho_P^r = \left(\pi_P^r, \beta_P^r \mid \delta_\Pi^I, \delta_P^I \right) \right)$ is repeated here for quick reference as in Figure 6.1. Big government means that the provision of goods and services in the hands of the public sector is relatively big by collective social assessment. Small government means that the provision of goods and services in the hands of the public sector is relatively small by collective social assessment. The concept of big or small as applied to any government is subjective and politically based within ideological confines of social thinking. In short, the Marxian political economy relates to labor control of the government and the means of production. The Schumpeterian political economy relates to capitalist control of the government and the means of production. These two form the extreme cases of the political economy of the private-public duality. In the

Marxian socialist political economy the three structures are public-sector-decision controlled, while in the Schumpeterian socialist political economy the three structures are private-sector-decision controlled. In the Schumpeterian political economy the social system is transformed into two power relations. There is the power of the people who controls nothing except their labor, where the peoples' power generates mass movements for socio-economic change. There is also the power of the bureaucratic-capitalists who control the economy generating resistance movement.

An important question tends to arise. How do we include the multitudes of governmental controls of private life in the size of the government? How do we include the social private regulations into the size of the government? Is a control of marriage relations or other social restrictions on private behavior included in the measurement of the size of the government? How do we relate government social activities and fiscal activities to the definition of the size of the government? A government may be small in fiscal activities and big in social activities as in the Zone IV. Similarly, a government may be big in fiscal activities and small in social activities in terms of control as in the Zone II. The government may be small in both fiscal and social activities as in the Zone I. Similarly, the government may be big in both fiscal and social activities as in the Zone III. These zones are presented in Figure 6.1.

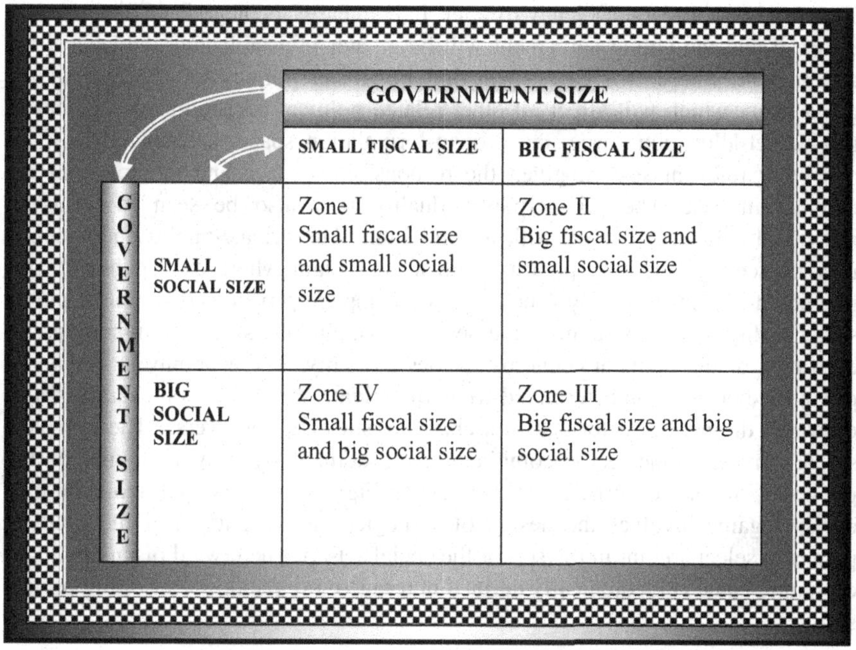

		GOVERNMENT SIZE	
		SMALL FISCAL SIZE	BIG FISCAL SIZE
G O V E R N M E N T S I Z E	SMALL SOCIAL SIZE	Zone I Small fiscal size and small social size	Zone II Big fiscal size and small social size
	BIG SOCIAL SIZE	Zone IV Small fiscal size and big social size	Zone III Big fiscal size and big social size

Fig. 6.1 The interactions between Fiscal and Social Size of Governments

The size of any government may be seen in terms of cost-benefit duality. The value of government's fiscal activities has value to the society as well as cost to the same society. Similarly, the value of government's social activities has value to the society as well as cost to the same society. Each of these zones is defined by two cost-characteristics sets and two-benefit-characteristic sets for measurements aggregation. It is common to think about the size of the government in terms of its cost characteristics set without reference to its benefit characteristics set. It is also common to think about the government size in terms of its fiscal-cost characteristics set without any reference to the cost characteristics set of social activities. The question is should the government size be measured as an absolute social cost size where the benefit to the society is of no interest. Alternatively should the government size be measured as relative social cost-benefit size where the social benefits and costs of the government are reconciled? It is possible that the cost-benefit approach to the measurement of the government may render traditionally viewed small government to be big.

Let us revisit the political duopoly in line with the political economy of the United States relative to the dynamics of private-public-sector duality. If the two parties agree on big government, they will be operating in Zone I of Figure 6.3, and hence, will negotiate on the size efficiency frontier above the 45^0-line in Figure 6.2. If they agree on small government, they will be operating in Zone III of Figure 6.3 and will negotiate for the size of government below the 45^0-line in Figure 6.2 on the size-efficiency frontier. In both cases the negotiated distance will be small and the voting agreement will be greater in establishing the size of the government. These two cases will find bipartisan politics on the size of the government which will affect all other related political decision-choice games in the budget-allocation space, the project-formulation space, accountability space and other relevant spaces, given the responsibility space and the space of the national interest. The private-public duality may also be seen in terms of individual-collective duality with a continuum. The private-public phenomenon may be seen in terms of polarity with a continuum where each pole contains individual-collective duality. The poles are complete private and complete public sectors while the continuum is the set of private-public-sector combinations. In each pole of the political economy resides a duality. The economy is said to be private-sector-decision determined when the private-sector decision characteristics outweigh the public-sector-decision characteristics and vice versa. The complete set of private-public-sector combinations is defined by moving points of the public-private sector efficiency frontier in Figure 6.2. The politico-economic decision game involves the design of strategies of negotiations of the political parties to select a point in the set for the social set-up. The reward of this politico-economic game is ideologically defined in terms of big public sector or big private sector.

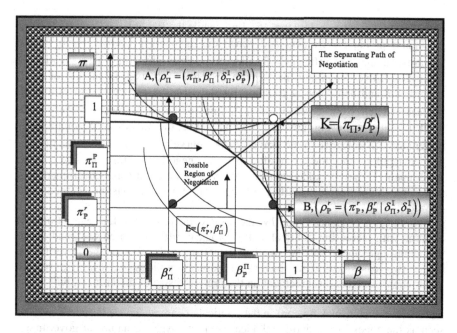

Fig. 6.2 The Negotiation Space and Path of Private-Public Sector Combination for Government Size

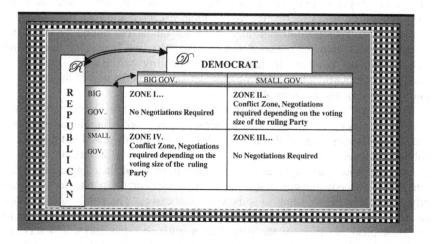

Fig. 6.3 The Small-Big Government Game and Negotiation Zones in the Political Economy

Intense problems of conflict arise, however, when the two parties find themselves in Zones II and IV where there are substantial disagreements over the preferred size of the government. In Zone II, the Democratic Party finds itself in support of small government and hence the members negotiate downwards on the size efficiency frontier below the 45^0-line, while the Republican Party finds itself in support of big government and hence the members negotiate its preference upwards on the size efficiency frontier above the 45^0-line. The situation is reversed in Zone IV, in that the Republican Party finds itself in support of small government and hence the negotiation for the size preference is downwards on the size efficiency curve below the 45^0-line, while the Democratic Party finds itself in support of big government and hence the members negotiate their preference size upwards on the size efficiency curve above the 45^0-line, the line of demarcation and the separating path of negotiation for the size of the government. It may be added that the preference for small (big) government is a preference for a bigger (smaller) private sector in the provision of goods and services.

In both cases of Zones II and IV, the disagreement distance for negotiation will be large with increasing political and ideological difficulty to close it, and hence the voting agreement will be small in establishing the size of the government. These two cases will find increasing partisan politics on the size of the government, and this disagreement on the preferred size of government will affect all related socio-political problems and the related sequence of political games within the decision-making core in the political economy. Zones II and IV may be seen as the zones of politico-ideological conflicts over the nature of government and governance, in that, the concept of big or small government is basically resolved in the ideological plane. These conflicts may be reduced by the presence of a party's super-majority in the decision-making core in which case the other party is reduced to an ineffective voting bloc, except if the minority party has available to it other institutional rules to use for sabotaging the decision-making system. It is within these two Zones that we will find increasing negotiation distrust, partisan conflicts in political accountability and political corruption, political witch-hunting, budgetary misconduct and intensification of the deceptive information structure to the public.

Zones I and III are politically peaceful but rarely occur. The people as the principal are lucky if they find their political economy operating in game zones I and III. Let us keep in mind that the size of the government viewed in terms of the public sector as well as the size of the private sector is defined in terms of provision of goods and services and measured in terms of their aggregate value. The size of the government is seen in terms of whether the public sector or the private sector is big or small in some relative measure. The public sector goods and services are provided by the government so that the size differences are due to the ideological positions of the parties in relation to what goods and services should the public sector supply, and what goods and services should the private sector supply in the political economy. The nature and structure of the provision of these goods and services are affected by the method of doing so. It may be pointed out that the debates on privatization decisions, government-outsourcing decisions and many such decisions are about the provision of goods and services, the nature of their provision, the method of their delivery and the distribution to the society in terms of who gets what, where and when relative to the provision of justice, opportunities, education, housing and many articles of social life. The conditions of the size preferences by different political parties are mapped onto different decision-choice games that are to be played by the members of the decision-making core through the negotiation functions and vote-casting processes. The conditions for the computation of the size of the government may be such that a political party may be in favor of small government fiscal activities and big government social activities and vice versa.

It may be pointed out that the justifications of preference for a government size usually proceed from two fronts of the political structure and the economic structure. At the level of economics, the justification is rooted in the efficiency criterion. This efficiency criterion is always twisted to favor the private sector and against the public sector in terms of the provision of goods and services in the nation. The construction of the private sector efficiency value for the supporting argument always neglects the effects of the elements of public-sector positive externalities, as well as its enhancements by the public sector social infrastructures and institutional support that enhance private-sector efficiencies. It also neglects the private-public sector relative benefits that are associated with the private-public relative size. Similarly, it neglects the negative private-sector externalities that ever-estimate the private sector efficiency by shifting costs to the public sector. This situation presents a case where the benefit-cost implication favors small government and the expansion of the private sector on the basis of distorted efficiency, where the cost of externalities is shifted to the public and benefits are reaped by the private sector. The implicit aggregate productivity value of government social services is not factored out off the private-sector productivity. In this way, the private sector efficiency norm is overestimated while the public sector efficiency norm is underestimated. From the complexity of the social system, the cost-benefit imputations of the control of the personal social life for stability by the governmental social activities are not included. What are the relative cost-benefit values of individual freedom, government social information restriction, freedom of assembly, free speech, lack of transparency and other things that we claim for democratic decision-choice system under the principle of majority rule?

At the level of politics, the efficiency criterion is replaced by an ideological criterion and the debate centers around capitalism-communism and hence around the best way of organizing the provision of goods and services to the society. Here, the general public operates on sentiments, ignorance, ideology and propaganda. At the level of natural rights, the debate centers on the best welfare social configuration through the best private-public sector combination for the provision of goods and services and their distribution in the political economy. There is no one single private-public sector combination that satisfies all societies with their culture, national interest and social vision. Similarly, even given one society there is no one single priva-public-sector combination that will satisfy all generations with changing preferences and cultural dynamics. Each society must solve its private-public sector combination problem that is appropriate for its cultural confines, national interest and social vision. Each generation, given the same society, must solve its private-public sector combination problem that is appropriate for its generational cultural confines. Let us turn our attention to a sequence of decision-choice games that is played by the members of the decision-making core in support of their size preferences in the political economy. Among these games are the credibility and trust of political negotiation, political accountability, political corruption and labor-capital support

6.2.2 The Good-Bad Faith Duality, Political Negotiations and the Voting Decisions

The democratic decision-choice process becomes more complicated by two sequential decision-choice modes of determining the size of the government and the distribution of the government budget. We must keep in mind that the budgetary decisions are entangled with the elements in the goal objective set and the program set, all of which are defined in the ideological space of the social dynamics. It is after the decision-choice action on the size of the government has been done that a good-bad faith game arises. From the good-bad faith, a political distrust is established in the democratic collective decision-choice process at the level of the social decision-making core. It may be pointed out that the democratic collective decision-choice process is as good as its information support. Ignorance and information deception (referred to in these discussions as *deceptive information structure*) lead to a mockery of the democratic process in the collective decision space. The negotiation game is like the prisoner's dilemma where the negotiation processes are then mapped into the trust space of good-bad duality of faith with a fuzzy continuum between good and bad. The game is to manipulate the relevant information to direct the opponent to undertake a sub-optimal action which the opponent would not have selected if the appropriate information was available. In this respect, the reward (benefit) of the game is the success of the information deception on the decision-choice action, and the cost of the game is the resource devoted to create the deceptive information structure. As such,, the negotiated residual of trust is the weighted distances of the individual good-bad distances called the *fuzzy political residual*. The size of the fuzzy political residual is defined in terms of weighted individual good-faith residuals and determined by the relative negotiation strength of the parties. The zones of trust are shown in Figure 6.4 for the case of a political duopoly that may be looked at from the structure of the political economy of the United States of America.

From the viewpoint of the general welfare and the efficiency of the democratic collective decision-choice system in the social formation, it would be socially enhancing if all political parties operate in Zone I where there is dual good faith in the negotiated decision-choice process. As we have explained, Zone III creates political decision- impasse and general public distrust in the governmental machinery and hence is not a good zone for the parties to operate in. However, the parties are forced by party affiliation and interests to operate more in this zone than is necessary. The existence of possible political gains and possible increased party welfare are motivations for the parties to operate in Zones II and IV where there is asymmetry of trust in negotiations. These zones may produce temporary politico-economic equilibria in favor of one party. The temporary possible

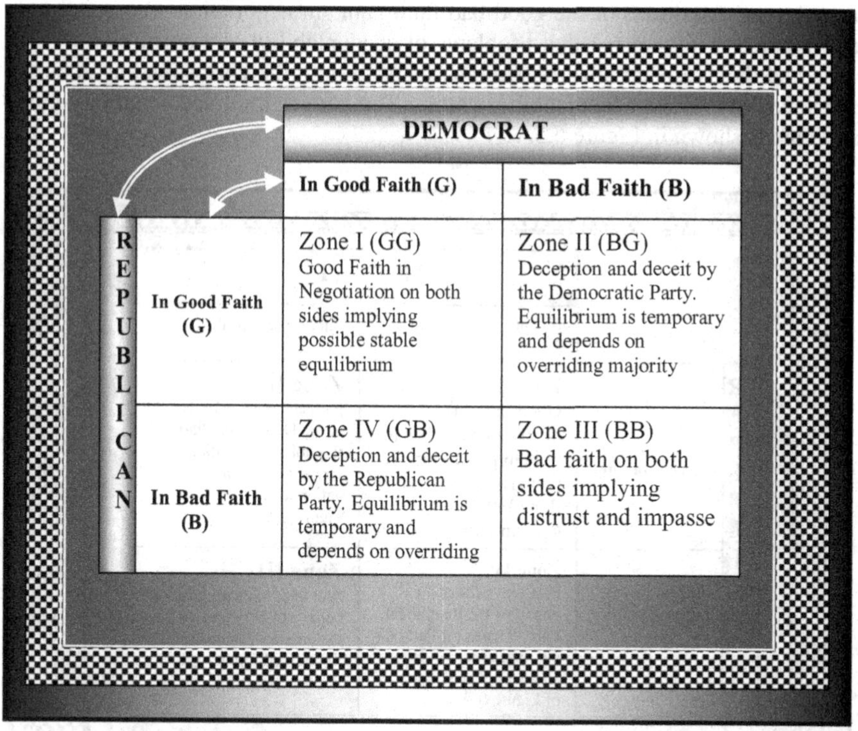

		DEMOCRAT	
		In Good Faith (G)	In Bad Faith (B)
R E P U B L I C A N	In Good Faith (G)	Zone I (GG) Good Faith in Negotiation on both sides implying possible stable equilibrium	Zone II (BG) Deception and deceit by the Democratic Party. Equilibrium is temporary and depends on overriding majority
	In Bad Faith (B)	Zone IV (GB) Deception and deceit by the Republican Party. Equilibrium is temporary and depends on overriding	Zone III (BB) Bad faith on both sides implying distrust and impasse

Fig. 6.4 The Good-Bad Faith Game in Negotiations in the Political Economy

decision-choice equilibrium in Zones II and IV of Figure 6.1 will depend on the relative size of the majority party and the relative size of the socio-economic power that the parties have to enforce their will. These zones are then mapped onto Zones II and IV of the accountability game whose structure is presented in Figure 6.5.

6.2.3   Party-Nation Duality and the Political Accountability Game in the Political Economy

Zone I of the bad-good faith of Figure 6.4 is mapped onto Zone I of Figure 6.5 of the accountability game. Zone III is unstable and cannot be maintained since it leads to a permanent collapse of governance with an increasing social cost that reduces the net benefit of governance to both the principal and the agent in the governing process. This zone is extremely partisan in the socio-political economy that creates popular distrust in the overall governmental institutional structure and operations leading to a blame game which cannot be sustained for a long political period. Zone III of Figure 6.4 is mapped onto Zone III of Figure 6.5 of the party-nation accountability game. This game involves the accountability position of the members of the decision-making core to either the nation or party of affiliation. It is under the conditions of the good-bad faith game in negotiation whose structure is presented in Figure 6.4 that problems of accountability and corruption tend to arise and their respective zones become mapped onto the corresponding zones of accountability of the accountability game. The party-nation accountability-game structure is shown in Figure 6.5.

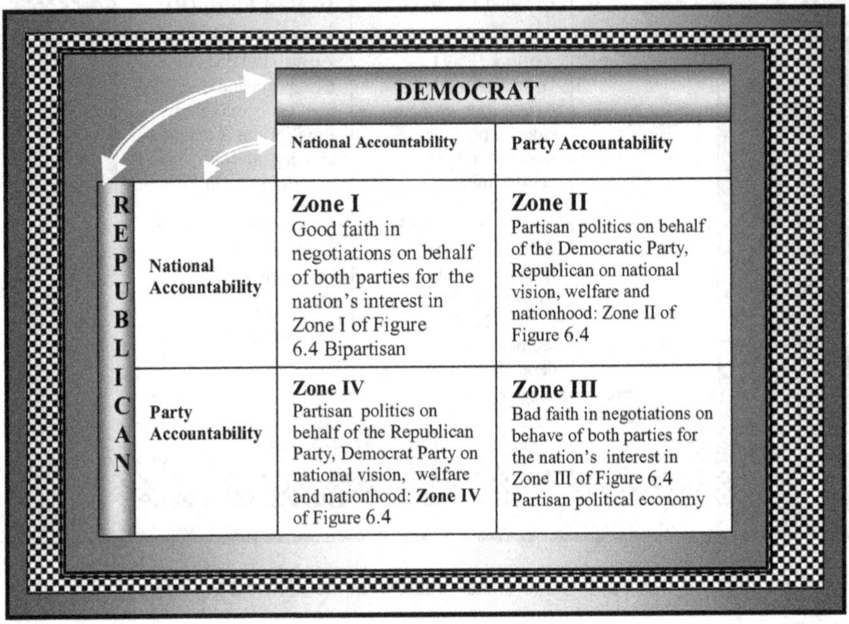

		DEMOCRAT	
		National Accountability	Party Accountability
R E P U B L I C A N	National Accountability	**Zone I** Good faith in negotiations on behalf of both parties for the nation's interest in Zone I of Figure 6.4 Bipartisan	**Zone II** Partisan politics on behalf of the Democratic Party, Republican on national vision, welfare and nationhood: Zone II of Figure 6.4
	Party Accountability	**Zone IV** Partisan politics on behalf of the Republican Party, Democrat Party on national vision, welfare and nationhood: **Zone IV** of Figure 6.4	**Zone III** Bad faith in negotiations on behave of both parties for the nation's interest in Zone III of Figure 6.4 Partisan political economy

Fig. 6.5 The Party-Nation Accountability Game in Political Negotiations in Decision-choice Actions in the Political Economy

Zone I, brings the members of the decision-making core together for the national interest and a reasonable structure of the common vision of the nation. In this zone there is less conflict on the constituent elements of the goal-objective set in support of the national interest and vision. Similarly, there is less disagreement on the budget-allocation decisions. Zone III presents extreme ideological conflicts and disagreements on the basic fundamentals of the national interest and vision. The differences translate into differences of the essential structure of the supporting goal-objective elements in support of any national interest and social vision. Partisan politics assumes a destructive stage with a national soul searching. Zones II and IV involve situations where the national interest and vision are being made to conform to the positions of various parties but are constrained by the acknowledgement of other party in the duopolistic political organization of the social set-up. The outcomes of Zones II and IV will depend on the relative strength of the parties in appealing to the principal of the political economy. Disagreements in the party ideologies in the construction of the national interest, the supporting goal-objective set, the corresponding programs and the final budget when done in good faith is healthy in strengthening the democratic collective decision-choice system, on the basis of complex system of social cost-benefit dualities.

The zones of the accountability game are then mapped onto the respective zones of the *corruption game*. The mapping function must be seen in terms of interdependent socio-political actions that relate to the nature of governance on the part of the social decision-making core. The corruption zones, in which party members will fall, will depend on their relative assessments of the corresponding cost-benefit configurations and the socio-economic incentive structure that is built into the political economy. Let us keep in mind here that we are working with a political economy whose fundamental principle is individualism relative to the spaces of interests and freedoms around which an incentive structure may be constructed. The incentive structure may be positive as a reward or may be negative as a punishment or a combination of both. The positive incentive is to encourage behaviors favorable to the national interest while the negative incentive is to discourage some behaviors. A combination of both is to reward and punish either to change or maintain a behavior.

The zone in which a party will fall in the small-big government game will depend on the party ideology and platform that will shape the cognitive computations of relative importance of cost-benefit attributes and the socioeconomic incentive elements in the political economy. The cost-benefit configurations and the incentive structures may be composed of measurable and non-measurable attributes which will directly or indirectly influence the direction of collective preferences of the party leading to a particular game zone that the party will select. In all these games, the individual in the decision-making core is assumed to be elected to follow the national interest and social welfare of the state but guided by the party ideology and platform as we have discussed. All the incentive structure and accountability may be complicated by a hidden individual personal interest and agenda. In the Schumpeterian political economy, the concept of individualism is practiced in a social context as seen through the relationship between labor and capital in terms of income distribution, freedom, responsibility and accountability. Let us now turn our attention to examine labor-capital duality in a negotiation game where capital is represented by business and labor is represented by the actual and potential workers.

Let us keep in mind that the government is a permanent institution and the social decision-making core is a temporary institution. The administration and management of institutions of the legal, political and economic structures rest in the hands of the members of the social decision-making core who deliver governance. The objective is essentially to balance decision-choice impacts on capital and labor in order to create fairness and acceptable degree of equity for socio-economic and welfare improvements of the collective with a social stability. It happens so often that such institutional administration and management are carried on in favor of capital or labor depending on the composite aggregate preference of the social decision-making core and the ideology of governance in selecting a point from the set of points in the labor-business duality. This point selection becomes an accountability game in the political economy to which we turn our attention.

6.2.4 Labor-Business Duality and the Political Accountability Game in the Political Economy

The structure of nation-party duality and the accountability game addresses the accountability question in relation to the nation or the party affiliation in the democratic collective decision-choice system. This structure may be replaced with the structure of business-labor duality and the accountability game that addresses the question in relation to decisions in favor of labor or business. In this way, we have partitioned the population in the politico-economic space into capital and labor classes in line with the Marxian political economy but embedded in the Schumpeterian political economy. Let us remember that in the Schumpeterian political economy, the labor as an agent of change is reduced to nothingness in the sense of inaction. The entrepreneurial capitalism is the starting point of analysis. The transformation dynamics of the political economy is seen in terms of games among classes of capitalists in the two structures of entrepreneurial capitalism and bureaucratic capitalism. The structural dynamics of the political economy is represented by the behaviors of the capitalists with either the government playing a mediating role or being an active participant in the politico-economic game. We move from the initial game between the two classes of entrepreneurs with innovation and innovation investment on one hand and entrepreneurs with non-innovation and non-innovation investment that deplete innovation profits creating risks to innovation profits on the other hand. A second stage then emerges where bureaucratic capitalism replaces entrepreneurial capitalism to reduce capital risk and protect profits, shifting the game among capitalist classes into a new game between the bureaucratic capitalist class and the social decision-making core representing the state as it has been explained in Chapter Three of this monograph.

The current game is played to define laws, rules and regulations in the legal sector that constrains the parameters of labor-business behavior in the political economy. Here, both the business and private sector may be grouped as one while labor, consumers and the public may be combined as one in relations to legislative politics in the establishment of the rules and regulations in the legal structure in favor of one or the other to a define fairness demarcation of behavior of the two classes. The structure of the game is about political accountability on the part of the parties to either labor or business. The game space is presented in Figure 6.6. Zones I and III are of little interest since the parties have similar or the same orientation. Zones II and IV generate competition between labor and business as classes seeking to have an influence on the members of the social decision-making core. The acquired influence may come to shape the cost-benefit assessments as well as corrupt the democratic voting-decision process of the members in the decision-making core. It is here that class conflicts between capital and labor show themselves at the doorsteps of the social decision-making core. The power to acquire influence is obtained through an effective organization that can exert pressure through a negative or positive incentive structure or both. The bureaucratic capitalist class through business is always organized. It control the organized money and other resources. To compete for the influence, labor must be effectively organized to obtained organized people. The influence-acquisition system is reduced to a duality of *organized money* and *organized people*. It is also the existence of potential power to compete against bureaucratic organized money in the in the political economy that every legal and non-legal attempts are made by capital to stop labor from creating an effective organization that will constitute channels of organized people.

It is here that labor and business struggle to acquire influence from the members of the social decision-making core where such a core is considered as politically neutral in the administrative-managerial social space. It is also here that control of money becomes an important corrupting influence on the democratic collective decision-choice process in favor of those who have money and organization to corrupt the voting outcomes of the social decision-making core. In Zone II, the democrats support business and hence are accountable to the business interest and less accountable to the interest of labor, while the Republicans support labor's interests and hence are accountable to labor. The party positions and conditions are reversed in Zone IV. In all these cases of zonal analysis, how should this class accountability in addition to the other accountability games be related to the general political accountability and the integrity of the democratic collective-decision-choice system? This question leads us to examine the corruption game in the democratic collective decision-choice structure of the

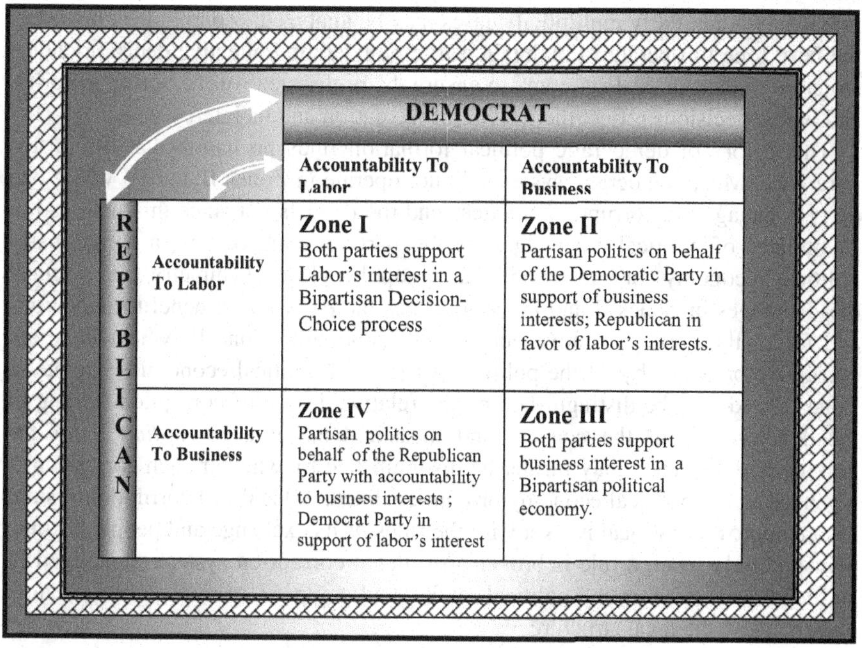

		DEMOCRAT	
		Accountability To Labor	Accountability To Business
R E P U B L I C A N	Accountability To Labor	**Zone I** Both parties support Labor's interest in a Bipartisan Decision-Choice process	**Zone II** Partisan politics on behalf of the Democratic Party in support of business interests; Republican in favor of labor's interests.
	Accountability To Business	**Zone IV** Partisan politics on behalf of the Republican Party with accountability to business interests ; Democrat Party in support of labor's interest	**Zone III** Both parties support business interest in a Bipartisan political economy.

Fig. 6.6 The Labor-Business Accountability Game of the Political Parties in the Political Economy

political economy. This is necessary because, it is at this competitive junction that political corruption emerges to make a mockery of democracy. In the case of Schumpeterian political economy of bureaucratic capitalism, where the three powers of economic, political and legal structures are controlled by the bureaucratic capitalist class, the accountability game is played in favor of capital in the capital-labor duality, where those who directly control capital services do not directly control the labor services. Those who directly control capital services indirectly control labor services through employment in the economic structure under the conditions of the legal structure that they have created from the political structure.

6.2.5 The Corruption Game, Political Market and the Political Economy

The corruption game is obtained by a decision mapping of the small-big government, party-nation accountability and labor-business accountability games into the legislative space where corruption tends to arise and a political market tends to be established. It is here that there arises an intense competition to acquire access and influence by manipulating the democratic collective decision-choice process through the purchasing of access and influence from members of the decision-making core by both business and labor. For analytical simplicity and other practical relevance, we shall consider a case of political duopoly as we have previously done. Party multiplicity cases may be analyzed. Zone I of Figure 6.6 is the best zone to operate to maintain a true democratic collective decision-choice structure in the political economy to ensure the preferences of the voting majority, and where decisions reflect the welfare of the nation and its interests.

The history of democratic political formation finds no nation that fits in this pure zone. Most democratic systems do not operate in Zones II and IV where one party is engaged in corruptive practices and the other is not since this will lead to asymmetry of political advantages on the part of some parties in a multiparty political economy or in a political duopoly. The asymmetry of political disadvantages in Zones II and IV complicates the party's cost-benefit imputations and forces the parties to move out of these zones into Zone III where there are corruptive practices by all the political parties. All political economies operate in Zone III and may be distinguished by the relative degree of corruption leading to the development of the concept and measurement of a *corruption index* for ranking nations. They may further be distinguished by whether such corruption is legalized in the political economy or non-legalized. A legalized corruption system has an approved political market with the rules of its exchange and people become punished only when a rule is broken. An illegal corruption system is operated in an illegal and underground political market and any participation in the system is punishable in the legal structure.

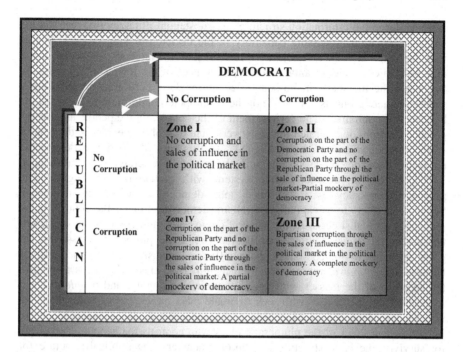

Fig. 6.7 The Party Corruption Game in the Political Economy with a Political Duopoly

The presence of legal corruption makes a mockery of democracy if democracy is seen in terms of a collective decision-choice system on the basis of the principle of majority rule without violence. The size of the corruption index obtained from the political market may be used to characterize the degree of mockery of democracy by the market system for the Schumpeterian political economy. The mockery of democracy is intensified by the presence of the deceptive information constraint through information restriction, disinformation and misinformation to the public by the established information disclosure rules, by any segment of the members of the social decision-making core on the principles of national security and public safety, and is further aggravated by a propaganda machine and an ideological war through effective mass media. The operation of the propaganda machine and ideological war in support of the mockery of democracy is such that those who seek to open the information space and report ill-doings are punished with stick, and the producers of the administrative evils as well as the violators of democratic principles are rewarded with carrots.

The characteristics of the deceptive information structure push the voters into sub-optimal decision space in the selection of the members of the social decision-making core, thus rendering the whole concept of democracy irrespective of how it is defined as meaningless. In these analyses, democracy, defined as an institutional arrangement to reconcile conflicts in the collective decision-choice space of social decision-choice elements without violence, fails if there is the presence of political markets for buying and selling of access and influence, extreme deceptive information structure and substantial illiteracy that restricts the social information processing capability in the decision-making process. In this way, the national goals and objectives and the national interests and social vision may be created to the advantage of the preferences of the few. This is especially the, when the members of the social decision-making core are selected from a small dominant group as in the case of the Schumpeterian political economy of the bureaucratic capitalism. The "government of the people, by the people and for the people" becomes degenerated to the "government to fool the people for the benefit of the privileged and of the privileged". In this case, the people's democratic collective decision-choice system with the people's sovereignty is transformed into a plutocratic decision-choice system with plutocratic sovereignty. The people's democracy is transformed into democratic plutocracy of which Schumpeter refers to as capitalist socialism (democracy) under the control of the plutocratic capitalist class. At the stage of democratic plutocracy of the social set-up, the social decision-choice conflicts between the plutocratic class and the government are dissolved. The social conflicts are transformed into the conflicts between *money power* represented by the plutocratic class who seeks to control the behavior of the masses with the plutocratic sovereignty, and *people's power* represented by the masses who seek to create mass-movement to claim their lost sovereignty. Here, the social battles in the collective decision-choice space are intense and in favor of the plutocratic class who controls the economic structure for survival, the legal structure and enforcement and the political structure for altering socio-political conditions.

6.3   Penumbral Regions of Rent-Seeking Decisions

We shall now examine the penumbral regions of rent-seeking activities in relation to the behavior of decision agents in the political market. The objective in this respect is to examine the utility in using the toolbox of the fuzzy paradigm to bring some useful understanding to the market decision-choice activities in the political markets. The fuzzy paradigm with it mathematics and logic is composed of a number of key concepts, analytical structures and computational techniques developed around principles of opposites, polarity, duality, continuum and the laws of thought [R7.14][R7.16][R15.13][R7.14]. This chapter is not about modeling and computational techniques. It is used to present a conceptual structure for the application of the technique and methods of fuzzy decisions to the analysis of the rent-seeking decisions in the political economy.

6.3.1   Fuzzy Decisions, Rent-Seeking and the Social Goal-Objective Formation

We shall now examine the nature of rent-seeking and the structure of fuzzy environments in the Schumpeterian political economy of bureaucratic capitalism. The penumbral regions of rent-seeking decision-choice activities are characterized by a fuzzy-stochastic environment. The fuzzy-stochastic environment is characterized by a defective and deceptive information structure. Decision making in the fuzzy environment requires the use of the framework of the principle of fuzzy rationality and how the fuzzy rationality affects the value computations of the social costs and benefits that will allow the integration of both the qualitative and quantitative characteristics of outcomes [R7.14][R15.13]. In line with the nature of the social information structure, we may cognitively question as to how fuzziness emerges out of the concept of political markets. We may then examine the conditions under which rent-seeking activities relate to the relevant fuzzy concepts and fuzzy rationality. The understanding of this question and the conditions of rent-seeking activities under fuzzy-stochastic information structure will lead us to the full appreciation of how fuzzy decisions and fuzzy equilibrium operate in political markets in relation to social goal-objective formations, project-selection decisions, and governmental budgetary decisions in relation to the general computational scheme of the social costs and benefits associated with various game decisions to reconcile collective conflicts in the political economy. Let us first look at: the fuzzy rationality and the value computations; political markets and fuzzy phenomenon; and rent-seeking and the fuzzy process, and then at the interaction of fuzzy decisions, political-market equilibria and choice of projects and funding. One of the important reasons for introducing the fuzzy approach into cost-benefit analysis is to broaden the domain of its applications to other areas of human decisions in addition to its usage in cases where markets, in a traditional sense, may not exist. This is the case when the variables in the decision system are overwhelmed by qualitative characteristics whose quantitative dispositions can be described in linguistic quantities such as big or small government.

6.3.1.1 The Concept of Fuzzy Rationality and Value Computations

The concept of fuzzy rationality may be viewed as a system of approximate rules that leads to a crisp decision-choice action with defined conditionality in support of the degree of confidence attached to the decision in a quantitative space. It is a systematic explanation and abstraction of regularities of successful fuzzy decisions and how these regularities are translated as approximate reasoning into prescriptive rules of good decisions under an approximation process [R7.13] [R7.16] [R7.67]. Such approximate rules allow value computations in terms of crisp costs and benefits. Fuzzy rationality also involves a critical investigation into the meaning and relevance of prescriptive approximate rules of decision as well as the development of consistent fuzzy algorithms to create crisp rules for practice in both fuzzy and fuzzy-stochastic environments. These rules are systematized into fundamental calculus for choosing alternatives in a fuzzy-decision environment through the methods of subjectivity, quantitative approximations and linguistic vagueness where the general information structure may be defective and also deceptive, thus increasing the qualitative component of the information set required for socio-political decisions in the political economy.

The approximate rules derived from the fuzzy-decision process constitute the intelligence of decision that allows crisp computations of cost-benefit values in a fuzzy environment. The computed values have a subjective embodiment in terms of human interpretation of the nature of the fuzzy environment and the accepted interpretations of the qualitative characteristics, especially when the information set is contaminated by deception or the information source is questionable. The computed actual and potential values include subjective costs and benefits that are invariably encountered in social and private choices involving the process of multi-criteria and multi-attribute decision situations. The structure of the subjective phenomenon, in the interpretation of the credibility of the information, is incorporated into the fuzzy rationality and then extended into cost-benefit analysis that allows elastic applications of the methods and techniques of cost-benefit analysis into decisions in political markets where social goals and objectives are determined and crystallized by legal precepts in a qualitative-quantitative space.

6.4 Political Markets and the Fuzzy Phenomenon in the Political Economy

Political markets, in the best democratic way, must reflect the model of perfectly competitive markets in the true economic sense that satisfies the postulates of perfect and unrestricted information flows, and atomicity without political monopolies and competitive imperfections. However, there are a number of elements that prevent the ideal perfect competition in the political markets. They include 1) the control of information flow by the members of the social decision-making core and its agencies by appealing to national security, public safety and others; 2) the existence, presence and active operations of mega political interest groups with differential resources to influence the outcomes of the prices of social goals and objectives; 3) ideological distortions of the channels of communication through deceptive information; 4) legal constraints that restrict people's behavior in the markets through the endorsement of political oligopolies, duopolies and other imperfections as well as outright exclusion of other social segments from political participation; 5) illiteracy and cognitive constraints in terms of bounded rationality; 6) the structure of income distribution that constrains effective participation; and 7) limitation rules that shrink the democratic participation space.

The effectiveness, integrity and efficiency of any democratic system of social decision-making depend on the genuine democratization of social information and knowledge, without which any talk of democracy representing in socio-economic decision-making is a fiction and ideological deception in the political economy. *Democratization of social information* is a vehicle to implement the postulate of perfect information regarding the articles of social choice. Democratization of education and learning is an important vehicle to relax the cognitive limitations of the decision-choice agents. Social decision-choice activities in the democratic collective-decision-choice space imply that the members of the decision-making core (the Government) should be transparent, and not operate in secrecy and classified information under the veil of national security regarding social decisions on any basis. Similarly, the channels of information production and dissemination must not restrict access or promote deception and misinformation on the basis of falsehood to influence social choices. Democratization of knowledge implies that the decision-making system must establish educational facilities that offer equal access to all members of the society. Participation in the social decision-making process must not be restricted by law and suppressive factors such as intimidation and unnecessary participation conditionality.

There are other factors that can either enhance or restrict the efficiency of the democratic collective decision-choice system. Increasing efficiency requires that there should not be the presence of mega groups whose actions impose pressures on the decision-making process to influence decision outcomes or generate conditions of non-wealth creating rents and revolving doors of favors and kickbacks. The presence of pressure groups creates imperfections in the political market and alters the market perfection into oligopolistic political markets. The possibility of creating a perfectly competitive political market is influenced by of think-tank institutions, the education process, ideological distortions that affect the collective social information structure. Each one of these has a role to play in creating either negative or positive effects on the outcomes of the democratic collective decision-choice system. The positive role of think-tanks in the democratic collective decision-choice process is to provide objective social information and analysis pertaining to social good and not to expand the space of deceptive information structure. The role of the educational process, in societies organized on the principles of the democratic collective decision-choice system, is to instruct, educate and develop the cognitive creativity and power of individual thinking that relate to analysis, synthesis and decision-making. These educational characteristics require the development of the cognitive ability to efficiently process objective information correctly by recognizing social truth and distinguish it from information deception and distortions. This is in relation to the ideal that people act on what they think and not what reality connotes. In fact, people's social realities are shaped by their perceptions and mental attitudes where thoughts are created by them which then influence intentions as well as define their social realities Ideological distortions of social information and knowledge creation should be absent or reduced to a minimum, since these distortions corrupt the channels of information transmission, reception and processing and hence affect social decisions and choices that are undertaken by the individuals and the collective. Additionally, the role of education in this democratic social decision-choice process is to improve individual and collective personality, and to reduce human limitations for information processing, where such information processing leads to social decisions and choices. All these are characteristics that enhance national collective personality and good citizenship in a society organized under the ideal principles of democratic collective decision-choice system.

These requirements for an efficient operation of the political markets fail in all democratic societies in varying degrees if the efficient social decision-making is based on individual preferences and cost-benefit calculations that are guided by the *classical rationality* in terms of information and knowledge use. The environment of the political market is usually tinted with mega groups, ideological twists among others, which restrict efficient social decision-making to establish social goals and objectives leading to the creation of relevant supporting social programs. The distortions are driven by rent-seeking, rent-creation and rent-protection activities that reward a few individuals at the expense of an increasing cost to the society.

The environment for social decision-making on the basis of individual preferences is characterized by stochastic and fuzzy uncertainties. The stochastic uncertainty reflects a lack of information in general while the fuzzy uncertainty reflects the presence of inexactness, subjectivity, ill-definedness, vagueness and others like deceptive information structure as induced by the presence of pressure interest groups and others such as rent-seekers. The effects of coalitions on the decisions to construct the social goal-objective set depend on the degree of participation of the members in the construct of mega-groups. The increasing effects of the degree of participation lead to the formation of *fuzzy coalitions* which allow inter-coalition movements. The rewards of rent-seeking activities in the political decision-choice process are of fuzzy values in the sense that exact values cannot be known. The analytical difficulty in the formulation and synthesis of the decision-making in constructing the social goal-objective set that produces the social benefit-cost system is simply how to incorporate the presence of fuzziness as induced by various interest groups, stakeholders and constituencies in search of rent, where the rewards are of fuzzy values which present themselves in terms of linguistic values. It is precisely the presence of such fuzzy characteristics in the political environment for the social decision-making process why the technique and methods of the theory of fuzzy decisions may be introduced in the construction of national interests, social vision and social goal-objective set based on the preferences of interest groups, stakeholders, policy makers and the general public.

6.5 Rent-Seeking as a Fuzzy-Decision Process

We have argued that the process of obtaining and constructing a social goal-objective set which includes goals, objectives social vision and national interests takes place in an environment where all the information characteristics of fuzziness are present. Thus, rent-seeking that accompanies goal-objective formation is a fuzzy decision-process. It is a fuzzy process in the sense that the rent-seeker must identify the inexact social goals and objectives that provide a reasonable size of potential rent that would justify the commitment of private resources to the actualization of a desirable goal-objective set. The whole concept of big-small government is a linguistic variable existing in duality and continuum whose exact quantitative definition is subjective under the principle of opposites. This big-small phenomenon leads to the question of how big is big and how small is small. Viewed from small-big duality, the behaviors of the fuzzy membership characteristic functions are such that as the membership characteristic function representing big government falls, the membership characteristic function representing small government rises and vice-versa defined in relation to the provision of goods and services. Associated with each potential social goal, objective and defined national interest is a potential rent that gives an impetus to activate rent-seeking activities whether it is wealth or non-wealth creating. The impetus to rent-creation, rent-protection, rent-activation and rent-harvesting is amplified when the private sector is large and the political economy is organized around unregulated profit motives with an established political market under legalized corruption.

Similarly, the impetus to rent-creation, rent-protection, rent-activation and rent-harvesting is magnified when the public productive capital is under privatization. In fact, the argument in support of privatization is also an argument in support of creating potential rent-seeking environments for profit enhancement at the expense of the public. The implied potential rent can be abstracted by interest groups when particular goals and objectives are incorporated into the social goal-objective set with the private delivery system for the provision of social goods and services. In order for any potential rent, implied in the social goal-objective set, to offer a possibility of actualization, rent seekers must first devote resources to prospect it for rent-loaded objectives; secondly, work for the particular social goals and objectives to be incorporated into the social goal-objective set; thirdly activate the potential rent for abstraction; and fourthly protect the rent-abstraction process by protecting particular social goals and objectives in the social goal-objective set. Each of these steps has elements of fuzziness and all these steps constitute an enveloping of organic fuzzy process of rent-seeking activities.

6.6 Equilibrium in Political Markets, Choice of Projects and Budgetary Decisions

The above sections speak to the method of rent-seeking, rent-creation and rent-protection as a fuzzy decision-choice process. The fuzzy process is an enveloping of a sequence of fuzzy decisions that are guided by *fuzzy rationality* as we have defined and pointed out [R7.14] [R7.15]. When the selection of social goals and objectives are viewed in terms of a democratic process and modeled in terms of a fuzzy phenomenon, then the optimal social objective set abstracted from the political market is composed of temporary equilibrium values that establish a priority ranking of social goals and objectives that are fuzzy decision-induced in the collective decision-space.

Each goal or objective will be supported by a set of projects that would allow either the goal or the objective to be actualized. It is at this stage of actualizing an objective that individuals and groups can activate their rent-seeking activities to fruition under an appropriate politico-legal environment and a defined incentive structure. The whole process of rent-seeking, rent-creation and rent-preservation demands influence creation and tampering on the political process as well as on the social, decision-making core (Government). It is thus the process of actualizing either a social goal or objective through project implementation that define the conditions in which the potential rent associated with either a goal or an objective can be abstracted by an individual, a firm or a group. The rent-abstraction is temporary if it is not accompanied by rent-protection. In this regard, the process of activities of rent-seeking is continuous because the set of social goals and objectives is dynamic and continuously being constituted and reconstituted for sequential socio-political game as new social problems tend to arise. The socio-political games in creating the set of social goals, objectives and national interests may be modeled by the logic of fuzzy games [R5.1][R5.2][R5.3] [R5.5] [R5.7]. The analysis of coalition formation, pattern recognition of interest development around social goals and objectives as well as interest-group stratification may be done with fuzzy logical reasoning [R6.63] [R6.111] [R6.138]. The construct of the fuzzy game may be defined in terms of fuzzy coalition or fuzzy payoff values where the payoff values may be related to the state of nature and probabilistically embedded with both fuzzy and stochastic conditionalities.

References: Interdisplanary

R1 Classical Game Theory, Information, and Decision-Choice Conflicts

[R1.1]   Aumann, R.: Correlated Equilibrium as an Expression of Bayesian Rationality. Econometrica 55, 1–18 (1987)

[R1.2]   Aumann, R., Hart, S.: Handbook of Game Theory with Economic Applications. North-Holland, New York (1992)

[R1.3]   Border, K.: Fixed Point Theorems with Applications to Economics and Game Theory. Cambridge University Press, Cambridge (1985)

[R1.4]   Brams, S., Kilgour, D.M.: Game Theory and National Security. Basil Blackwell, Oxford (1988)

[R1.5]   Brandenburger, A.: Knowledge and Equilibrium Games. Journal of Economic Perspectives 6, 83–102 (1992)

[R1.6]   Campbell, R., Sowden, L.: Paradoxes of Rationality and Cooperation: Prisoner's Dilemma and Newcomb's Problem. University of British Columbia Press, Vancouver (1985)

[R1.7]   Crawford, V., Sobel, J.: Strategic Information Transmission. Econometrica 50, 1431–1452 (1982)

[R1.8]   Dresher, M., et al. (eds.): Contributions to the Theory of Games. Annals of Mathematics Studies, No. 39, vol. III. Princeton University Press, Princeton (1957)

[R1.9]   Scott, G., Humes, B.: Games, Information, and Politics: Applying Game Theoretic Models to Political Science. University of Michigan Press, Ann Arbor (1996)

[R1.10]   Gjesdal, F.: Information and Incentives: The Agency Information Problem. Review of Economic Studies 49, 373–390 (1982)

[R1.11]   Harsanyi, J.: Games with Incomplete Information Played by 'Bayesian' Players I: The Basic Model. Management Science 14, 159–182 (1967)

[R1.12]   Harsanyi, J.: Games with Incomplete Information Played by 'Bayesian' Players II: Bayesian Equilibrium Points. Management Science 14, 320–334 (1968)

[R1.13]   Harsanyi, J.: Games with Incomplete Information Played by 'Bayesian' Players III: The Basic Probability Distribution of the Game. Management Science 14, 486–502 (1968)

[R1.14]   Harsanyi, J.: Rational Behavior and Bargaining Equilibrium in Games and Social Situations. Cambridge University Press, New York (1977)

[R1.15]   Haussmann, U.G.: A Stochastic Maximum Principle for Optimal Control of Diffusions. Longman, Essex (1986)

[R1.16]   Haywood, O.: Military Decisions and Game Theory. Journal of the Operations Research Society of America 2, 365–385 (1954)

[R1.17]   Krasovskii, N.N., Subbotin, A.I.: Game-theoretical Control Problems. Springer, New York (1988)

[R1.18]   Kuhn, H. (ed.): Classics in Game Theory. Princeton University Press, Princeton (1997)

[R1.19]   Lagunov, V.N.: Introduction to Differential Games and Control Theory. Heldermann Verlag, Berlin (1985)

[R1.20]   Luce, D.R., Raiffa, H.: Games and Decisions. John Wiley and Sons, New York (1957)

[R1.21]   Maynard Smith, J.: The Theory of Games and the Evolution of Animal Conflicts. Journal of Theoretical Biology 47, 209–221 (1974)

[R1.22]   Maynard Smith, J.: Evolution and the Theory of Games. Cambridge University Press, Cambridge (1982)

[R1.23]   Milgrom, P., Roberts, J.: Rationalizablility, Learning and Equilibrium in Games with Strategic Complementarities. Econometrica 58, 1255–1279 (1990)

[R1.24]   Myerson, R.: Game Theory: Analysis of Conflict. Harvard University Press, Cambridge (1991)

[R1.25]   Nalebuff, B., Riley, J.: Asymmetric Equilibrium in the War of Attrition. Journal of Theoretical Biology 113, 517–527 (1985)

[R1.26]   Pauly, M.V.: Clubs Commonality, and the Core: An Integration of Game Theory and the Theory of Public Goods. Economica 35, 314–324 (1967)

[R1.27]   Rapoport, A., Chammah, A.: Prisoner's Dilemma: A Study in Conflict and Cooperation. University of Michigan Press, Ann Arbor (1965)

[R1.28]   Roth, A.E.: The Economist as Engineer: Game Theory, Experimentation, and Computation as Tools for Design Economics. Econometrica 70, 1341–1378 (2002)

[R1.29]   Shubik, M.: Game Theory in the Social Sciences: Concepts and Solutions. MIT Press, Cambridge (1982)

[R1.30]   Von Neumann, J., Morgenstern, O.: The Theory of Games in Economic Behavior. John Wiley and Sons, New York (1944)

R2 Collective Rationality, Public Choice and Social Decision-Choice Process

[R2.1]   Aaron, H.J., et al.: Efficiency and Equity in the Optimal Supply of a Public Good. Rev. of Econ. Stat. 51, 31–39 (1969)

[R2.2]   Aaron, H.J., et al.: Public Goods and Income Distribution. Econometrica 38, 907–920 (1970)

[R2.3]   Apter, D.E.: Choice and politics of allocation. Yale University Press, New Haven (1971)

[R2.4]   Archibald, G.C.: Welfare Economics Ethics and Essentialism. Economica (New Series) 26, 316–327 (1959)

[R2.5]   Arrow, K.J.: Social Choice and Individual Values. Wiley, New York (1951)

[R2.6]   Arrow, K.J.: Behavior under Uncertainty and Its Implication for Policy. Technical Report 399, Center for Research on Organizational Efficiency, Stanford, Stanford University (1983)

[R2.7]   Arrow, K.J.: A Difficulty in the Concept of Social Welfare. Jour. of Political Econ. 58, 328–346 (1950)

[R2.8]   Arrow, K.J.: Equality in Public Expenditure. Quarterly Jour. of Econ. 85, 409–415 (1971)

[R2.9]   Arrow, K.J., et al. (eds.): Readings in Welfare Economics. Homewood,
            Irwin (1969)
[R2.10]   Atkinson, A.B., et al.: Lecture on Public Economics. McGraw-Hill, New
            York (1980)
[R2.11]   Atkinson, A.B.: Optimal Taxation and the Direct Versus Indirect Tax
            Controversy. Canadian Jour. of Economics 10, 590–606 (1977)
[R2.12]   Atkinson, A.B., et al.: The Structure of Indirect Taxation and Economic
            Efficiency. Jour. of Public Economics 1, 97–119 (1972)
[R2.13]   Atkinson, A.B.: The Design of Tax Structure: Direct Versus Indirect
            Taxation. Journal of Public Econ. 6, 55–75 (1977)
[R2.14]   Axelrod, R.: Conflict of Interest, Chicago, Markham (1970)
[R2.15]   Balassa, B., et al.: Economic Progress, Private Values and Public Policy.
            North-Holland, New York (1977)
[R2.16]   Baldwin, D.A.: Foreign Aid, Interventions and Influence. World Politics 21,
            425–447 (1969)
[R2.17]   Baumol, W.J.: Welfare Economics and the Theory of the State. Harvard
            University Press, Cambridge (1967)
[R2.18]   Becker, G.S.: A Theory of Competition among pressure Groups for
            Political Influence. Quarterly Jour. of Econ. 97(XCVII), 371–400 (1983)
[R2.19]   Becker, G.S.: A Theory of Social Interactions. Jour. of Political
            Economy 82, 1063–1093 (1974)
[R2.20]   Bergson, A.: A Reformation of Certain Aspects of Welfare Economics.
            Quart. Jour. of Econ. 52, 314–344 (1938)
[R2.21]   Bergson, A.: On the Concept of Social Welfare. Quarterly Jour. of
            Econ. 68, 233–253 (1954)
[R2.22]   Bernholz, P.: Economic Policies in a Democracy. Kyklos 19(fasc. 1), 48–80
            (1966)
[R2.23]   Black, D.: On the Rationale of Group Decision Making. Jour. of Political
            Economy 56, 23–34 (1948)
[R2.24]   Blackorby, C., et al.: Utility vs. Equity. Jour. Public Economics 7, 365–381
            (1977)
[R2.25]   Blau, J.H., et al.: Social Decision Functions and the Veto. Econometrica 45,
            871–879 (1977)
[R2.26]   Bowen, H.R.: The Interpretation of Voting in the Allocation of Economic
            Resources. Quart. Jour. of Econ. 58, 27–48 (1943)
[R2.27]   Brown, D.J.: Aggregation of Preferences. Quart. Jour. of Econ. 89, 456–469
            (1975)
[R2.28]   Buchanan, J.M.: Individual Choice in Voting and the Market. Jour. of
            Political Econ. 62, 334–343 (1954)
[R2.29]   Buchanan, J.M.: An Economic Theory of Clubs. Economica 32, 1–14
            (1965)
[R2.30]   Buchanan, J.M.: Notes for an Economic Theory of Socialism. Public
            Choice 8, 29–43 (1970)
[R2.31]   Buchanan, J.M.: The Demand and Supply of Public Goods. Rand McNally,
            Chicago (1968)
[R2.32]   Buchanan, J.M.: Public Finance and Public Choice. National Tax Jour. 28,
            383–394 (1975)
[R2.33]   Buchanan, J.M., et al. (eds.): Theory of Public Choice. The University of
            Michigan Press, Ann Arbor (1972)
[R2.34]   Buchanan, J.M., et al. (eds.): Toward a Theory of the Rent Seeking Society,
            College Station. Texas A and M Univ. Press, Texas (1980)

[R2.35]   Buchanan, J.M., et al.: The Calculus of Consent. The University of
            Michigan Press, Ann Arbor (1962)
[R2.36]   Chipman, J.S.: The Welfare Ranking of Pareto Distributions. Jour. of Econ.
            Theory 9, 275–282 (1974)
[R2.37]   Coleman, J.S.: Foundations for a Theory of Collective Decisions. Amer.
            Jour. of Sociology 71, 615–627 (1966)
[R2.38]   Coleman, J.S.: The Possibility of a Social Welfare Function. Amer. Econ.
            Rev. 56, 1105–1112 (1966)
[R2.39]   Coleman, J.S.: The Possibility of a Social Welfare Function: A Reply.
            Amer. Econ. Rev. 57, 1311–1317 (1967)
[R2.40]   Comanor, W.S.: The Median Voter and the Theory of Political Choice.
            Jour. of Public Econ. 5, 169–177 (1976)
[R2.41]   Corlette, W., et al.: Complementality and Excess Burden of Taxation. Rev.
            Econ. Studies 21, 21–30 (1954)
[R2.42]   D'Aspermont, C., et al.: Equity and the Information Basis of Collective
            Choice. Rev. of Econ. Stud. 44, 119–209 (1977)
[R2.43]   DeMeyer, F., et al.: A Welfare Function Using Relative Intensity of
            Preferences. Quart. Jour. of Econ. 85, 179–186 (1971)
[R2.44]   Diamond, P.: Cardinal Welfare, Individualistic Ethics, and Interpersonal
            Comparisons of Utility: A Comment. Jour. of Political Econ. 75, 765–766
            (1967)
[R2.45]   Dobb, M.A.: Welfare Economics and the Economics of Socialism.
            Cambridge Univ. Press, Cambridge (1969)
[R2.46]   Downs, A.: Economic Theory of Democracy. Harper and Row, NewYork
            (1957)
[R2.47]   Dubins, L.E., et al.: How to Cut a Cake Fairly. Amer. Math Monthly, 1–17
            (1968); Reprinted in Newman, P. (ed.): Readings in Mathematical
            Economics. Value Theory, vol. 1. John Hopkins Univ. Press, Baltimore
            (1968)
[R2.48]   Ellickson, B.: A Generalization of the Pure Theory of Public Goods.
            American Econ. Rev. 63, 417–432 (1973)
[R2.49]   Farquharson, R.: Theory of Voting. Yale University Press, New Haven
            (1969)
[R2.50]   Fontaine, E.: Economic Principles for Project Evaluation. Organization of
            American States, Washington, DC (1975)
[R2.51]   Green, J., Laffont, J.-J.: Characterization of Satisfactory Mechanisms for
            Revelation of Preferences for Public Goods. Econometrica 45, 427–438
            (1977)
[R2.52]   Green, J., et al.: Imperfect Personal Information and the Demand Revealing
            Process: A Sampling Approach. Public Choice 29, 79–94 (1977)
[R2.53]   Green, J.R., et al. (eds.): Incentives in Public Decision-Making. North-
            Holland, New York (1979)
[R2.54]   Haefele, E.T.: A Utility Theory of Representative Government. Amer.
            Econ. Rev. 61, 350–367 (1971)
[R2.55]   Hammond, P.J.: Why Ethical Measures of Inequality Need Interpersonal
            Comparisons. Theory and Decision 7, 263–274 (1976)
[R2.56]   Harberger, A.C.: The Basic Postulates for Applied Welfare Economics; An
            Interpretive Essay. Jour. of Econ. Lit. 9(3), 785–793 (1971)
[R2.57]   Hardin, R.: Collective Action as an Agreeable n-Prisoner's Dilemma.
            Behav. Science 16, 471–481 (1971)
[R2.58]   Hause, J.C.: The Theory of Welfare Cost Measurement. Jour. of Polit.
            Econ. 83, 1145–1182 (1975)

[R2.59]   Head, J.G., et al.: Public Goods, Private Goods and Ambiguous Goods. Econ. Journal 79, 567–572 (1969)

[R2.60]   Heineke, J.M.: Economic Model of Criminal Behavior. North-Holland, New York (1978)

[R2.61]   Hicks, J.R.: The Four Consumer's Surpluses. Rev. of Econ. Stud. 13, 68–73 (1944)

[R2.62]   Hicks, J.R.: The Foundations of Welfare Economics. Economic Journal 59, 696–712 (1949)

[R2.63]   Intrilligator, M.D.: A Probabilistic Model of Social Choice. Rev. Econ. Stud. 40, 553–560 (1973)

[R2.64]   Jakobsson, U.: On the Measurement of the Degree of Progression. Jour. of Public Economics 14, 161–168 (1976)

[R2.65]   Kaldor, N.: Welfare Propositions of Economics and Interpersonal Comparison. Economic Jour. 49, 549–552 (1939)

[R2.66]   Kaldor, N. (ed.): Conflicts in Policy objectives. Oxford Univ. Press, Oxford (1971)

[R2.67]   Kemp, M.C., et al.: More on Social Welfare Functions: The Incompatibility of Individualism and Ordinalism. Economica 44, 89–90 (1977)

[R2.68]   Kemp, M.C., et al.: On the Existence of Social Welfare Functions: Social Orderings and Social Decision Function. Economica 43, 59–66 (1976)

[R2.69]   Kregel, J.A.: The Reconstruction of Political Economy: An Introduction to Post-Keynesian Economics. John Wiley and Sons, New York (1973)

[R2.70]   Laffont, J.-J. (ed.): Aggregation and Revelation of Preferences. North-Holland, New York (1979)

[R2.71]   Lancaster, K.J.: A New Approach to Consumer Theory. Jour. of Political Economy 74, 132–157 (1966)

[R2.72]   Lancaster, K.J.: Variety, Equity, and Efficiency. Columbia University Press, New York (1979)

[R2.73]   Lin, S.A.Y. (ed.): Theory and Measurement of Economic Externalities. Academic Press, New York (1976)

[R2.74]   Lindsay, C.M.: A Theory of Government Enterprise. Jour. of Political Econ. 84, 1061–1077 (1976)

[R2.75]   Lipsey, R.G., et al.: The General Theory of Second Best. Rev. of Econ. Stud. 24, 11–32 (1957)

[R2.76]   Little, I.M.D.: Social Choice and Individual Values. Jour. of Political Econ. 60, 422–432 (1952)

[R2.77]   Little, I.M.D.: A Critique of Welfare Economics. Clarendon Press, Oxford (1957)

[R2.78]   McFadden, D.: The Revealed Preferences of a Public Bureaucracy Theory. Bell Jour. of Econ. 6(2), 55–72 (1975)

[R2.79]   McFadden, D.: The Revealed Preferences of a Government Bureaucracy: Empirical Evidence. Bell Jour. of Econ. 7(1), 55–72 (1976)

[R2.80]   McGuire, M.: Private Good Clubs and Public Good Clubs: Economic Model of Group Formation. Swedish Jour. of Econ. 74, 84–99 (1972)

[R2.81]   McGuire, M.: Group Segregation and Optimal Jurisdiction. Jour. of Political Econ. 82, 112–132 (1974)

[R2.82]   Mirkin, B.: Group Choice. John Wiley, New York (1979)

[R2.83]   Millerson, J.C.: Theory of Value with Public Goods: A Survey Article. Jour. Econ. Theory 5, 419–477 (1972)

[R2.84]   Mishan, E.J.: Survey of Welfare Economics: 1939-1959. In: Surveys of Economic Theory, vol. 1, pp. 156–222. Macmillan, New York (1968)

[R2.85]   Mishan, E.J.: Welfare Criteria: Resolution of a Paradox. Economic Journal 83, 747–767 (1973)

[R2.86]   Mishan, E.J.: Flexibility and Consistency in Cost Benefit Analysis. Economica 41, 81–96 (1974)

[R2.87]   Mishan, E.J.: The Use of Compensating and Equivalent Variation in Costs-Benefit Analysis. Economica 43, 185–197 (1976)

[R2.88]   Mishan, E.J.: Introduction to Normative Economics. Oxford Univ. Press, New York (1981)

[R2.89]   Mueller, D.C.: Voting Paradox. In: Rowley, C.K. (ed.) Democracy and Public Choice, pp. 77–102. Basil Blackwell, New York (1987)

[R2.90]   Mueller, D.C.: Public Choice: A Survey. Jour. of Econ. Lit. 14, 396–433 (1976)

[R2.91]   Mueller, D.C.: The Possibility of a Social Welfare Function: Comment. Amer. Econ. Rev. 57, 1304–1311 (1967)

[R2.92]   Mueller, D.C.: Allocation, Redistribution and Collective Choice. Public Finance 32, 225–244 (1977)

[R2.93]   Mueller, D.C.: Voting by Veto. Jour. of Public Econ. 10, 57–75 (1978)

[R2.94]   Mueller, D.C.: Public Choice. Cambridge Univ. Press, New York (1979)

[R2.95]   Mueller, D.C., et al.: Solving the Intensity Problem in a Representative Democracy. In: Leiter, R.D., et al. (eds.) Economics of Public Choice, pp. 54–94. Cyro Press, New York (1975)

[R2.96]   Musgrave, R.A., et al. (eds.): Classics in the Theory of Public Finance. St. Martin's Press, New York (1994)

[R2.97]   Musgrave, R.A.: Public Finance in a Democratic Society: Collected Papers. New York Univ. Press, New York (1986)

[R2.98]   Newbery, D., et al. (eds.): The Theory of Taxation for Developing Countries. Oxford University Press, New York (1987)

[R2.99]   Olson, M.: The Logic of Collective Action. Harvard Univ. Press, Cambridge (1965)

[R2.100] Park, R.E.: The Possibility of a Social Welfare Function: Comment. Amer. Econ. Rev. 57, 1300–1304 (1967)

[R2.101] Pattanaik, P.K.: Voting and Collective Choice. Cambridge Univ. Press, New York (1971)

[R2.102] Pauly, M.V.: Cores and Clubs. Public Choice 9, 53–65 (1970)

[R2.103] Plott, C.R.: Ethics, Social Choice Theory and the Theory of Economic Policy. Jour. of Math. Soc. 2, 181–208 (1972)

[R2.104] Plott, C.R.: Axiomatic Social Choice Theory: An Overview and Interpretation. Amer. Jour. Polit. Science 20, 511–596 (1976)

[R2.105] Rae, D.W.: Decision-Rules and Individual Values in Constitutional Choice. Amer. Polit. Science Rev. 63, 40–56 (1969)

[R2.106] Rae, D.W.: The Limit of Consensual Decision. Amer. Polit. Science Rev. 69, 1270–1294 (1975)

[R2.107] Rapport, A., et al.: Prisoner's Dilemma. Michigan University Press, Ann Arbor (1965)

[R2.108] Rawls, J.A.: A Theory of Justice. Harvard Univ. Press, Cambridge (1971)

[R2.109] Rawls, J.A.: Concepts of Distributional Equity: Some Reasons for the maximum Criterion. Amer. Econ. Rev. 64, 141–146 (1974)

[R2.110] Ray, P.: Independence of Irrelevant Alternatives. Econometrica 41, 987–991 (1973)

[R2.111] Reimer, M.: The Case for Bare Majority Rule. Ethics 62, 16–32 (1951)

[R2.112] Reimer, M.: The Theory of Political Coalition. Yale Univ. Press, New Haven (1962)

[R2.113] Roberts, F.S.: Measurement Theory: with Applications to Decision Making, Utility and the Social Science. Addison-Wesley, Reading (1979)
[R2.114] Roberts, K.W.S.: Voting Over Income Tax Schedules. Jour. of Public Econ. 8, 329–340 (1977)
[R2.115] Rothenberg, J.: The Measurement of Social Welfare. Prentice-Hall, Englewood Cliffs (1961)
[R2.116] Satterthwaite, M.A.: Strategy-Proofness and Arrows Conditions: Existence and Correspondence Theorem for Voting Procedures and Social Welfare Functions. Jour. of Econ. Theory 10, 187–217 (1975)
[R2.117] Schneider, H.: National Objectives and Project Appraisal in Developing Countries. Development Centre of OECD, Paris (1975)
[R2.118] Sen, A.K.: A Possibility Theorem on Majority Decisions. Econometrica 34, 491–499 (1966)
[R2.119] Sen, A.K.: Quasi-transitivity, Rational Choice and Collective Decisions. Rev. Econ. Stud. 36, 381–394 (1969)
[R2.120] Sen, A.K.: Rawls versus Benthan: An Axiomatic Examination of the Pure Distribution Problem. Theory and Decision 4, 301–310 (1974)
[R2.121] Sen, A.K.: Informational Basis of Alternative Welfare Approaches Aggregation and Income Distribution. Jour. Public Econ. 3, 387–403 (1974)
[R2.122] Sen, A.K.: Liberty, Unanimity and Rights. Economica 43, 217–245 (1976)
[R2.123] Sen, A.K.: Social Choice Theory: A Re-examination. Econometrica 45, 43–89 (1977)
[R2.124] Sen, A.K.: On Weight and Measures Informational Constraints in Social Welfare Analysis. Econometrica 45 (October 1977)
[R2.125] Sen, A.K.: Collective Choice and Social Welfare. Holden-Day, San Francisco (1970)
[R2.126] Siegan, B.H.: Economic Liberties and the Constitution. Univ. of Chicago Press, Chicago (1980)
[R2.127] Stone, A.H., et al.: Generalized Sandwich Theorems. Duke Math. Jour. 9, 356–359 (1942)
[R2.128] Taylor, M.J.: Graph Theoretical Approach to the Theory of Social Choice. Public Choice 4, 35–48 (1968)
[R2.129] Taylor, M.J.: Proof of a Theorem on Majority Rule. Behavioral Science 14, 228–231 (1969)
[R2.130] Tideman, J.N., et al.: A New and Superior Process for Making Social Choice. Jour. of Polit. Economy 84, 1145–1159 (1976)
[R2.131] Tollison, R.M., et al.: Information and Voting: An Empirical Note. Public Choice 24, 43–49 (1975)
[R2.132] Tullock, G.: Some Problems of Majority Voting. Jour. of Polit. Econ. 67, 571–579 (1959)
[R2.133] Tullock, G.: The Politics of Bureaucracy. Public Affairs Press, Washington, DC (1965)
[R2.134] Varian, H.R.: Equity, Envy and Efficiency. Jour. Econ. Theory 9, 63–91 (1974)
[R2.135] Varian, H.R.: Two Problems in the Theory of Fairness. Jour. of Public Econ. 5, 249–260 (1976)
[R2.136] Williamson, O.E., et al.: Social Choice: A Probabilistic Approach. Econ. Jour. 77, 797–813 (1967)
[R2.137] Wilson, R.: A Game-Theoretic Analysis of Social Choice. In: Liebermann, B. (ed.) Social Choice, pp. 393–407. Gordon and Breach, New York (1971)
[R2.138] Wilson, R.A.: Stable Coalition Proposals in Majority-Rule Voting. Jour. of Econ. Theory 3, 254–271 (1971)

[R2.139]   Wingo, L., et al. (eds.): Public Economics and Quality of Life. Hopkins Univ. Press, John Baltimore (1977)
[R2.140]   Yaari, M.E., et al.: On Dividing Justly. Social Choice and Welfare 1, 1–24 (1984)

R3 Cost-Benefit Foundations for Decision-Choice System

R3.1 Cost-Benefit Rationality and Contingent Valuation Method (CVM)

[R3.1.1]   Ajzen, I., Fishbein, M.: Understanding Attitudes and Predicting Social Behavior. Prentice-Hall, Inc., Englewood Cliffs (1980)
[R3.1.2]   Arrow, K., et al.: Repeat of NOAA Panel on Contingent Valuation. Federal Register 58, 4601–4614 (1993)
[R3.1.3]   Bateman, I., Willis, K. (eds.): Valuing Environmental Preference: Theory and Practice of the Contingent Valuation Method in the US, EC and Developing Countries. Oxford University Press, Oxford (2002)
[R3.1.4]   Batie, S.S., et al.: Valuing Non-Market Goods-Conceptual and Empirical Issues: Discussion. Amer. Jour. of Agricultural Economics 61(5), 931–932 (1979)
[R3.1.5]   Bentkover, J.D., et al. (eds.): Benefit Assessment: The State of the Art. D.Reidel, Boston (1986)
[R3.1.6]   Bishop, R.C., Heberlein, T.A.: The Contingent Valuation Method. In: Johnson, R.L., Johnson, G.V. (eds.) Economic Valuation of Natural Resources: Issues, Theory, and Applications, pp. 81–104. Westview Press, Boulder (1990)
[R3.1.7]   Bishop, R.C., et al.: Contingent Valuation of Environmental Assets: Comparison with a Simulated Market. National Resources Jour. 23(3), 619–634 (1983)
[R3.1.8]   Brookshire, D.S., et al.: The Advantage of Contingent Valuation Methods for Benefit Cost Analysis. Public Choice 36(2), 235–252 (1981)
[R3.1.9]   Brookshire, D.S., et al.: Valuing Public Goods: A Compromise of Survey and Hedonic Approaches. Amer. Economic Review 72(1), 165–177 (1982)
[R3.1.10] Burness, H.S., et al.: Valuing Policies which Reduce Environmental Risk. Natural Resources 23(3), 675–682 (1983)
[R3.1.11] Carson, R.T., et al.: Temporal Reliability of Estimates from Contingent Valuations, Discussion Paper 95-37. Resource for the Future, Washington, DC (1995)
[R3.1.12] Carson, R., et al.: A Contingent Valuation Study of Lost Passive Use Values Resulting from the Exxon Valdex Oil Spill, Report to the Attorney General of Alaska. Natural Resource Damage Assessment, Inc., La Jolla (1992)
[R3.1.13] Cummings, R.G., Brookshire, D.S., Schulze, W.D., et al. (eds.): Valuing Environmental Goods: An Assessment of the Contingent Valuation Method. Rowman and Allanheld, Totowa (1996)
[R3.1.14] Diamond, P.A., Hausman, J.A.: On Contingent Valuation Measurement of Nonuse Values. In: Hausman, J. (ed.) Contingent Valuation: A Critical Assessment, pp. 3–38. North- Holland Press, Amsterdam (1993)

[R3.1.15]   Diamond, P.A., et al.: Contingent Valuation: Is Some Number Better then no Number? Jour. Economics Perspectives 8, 45–64 (1994)

[R3.1.16]   Dompere, K.K.: The Theory of Fuzzy Decisions, Cost Distribution Principle in Social Choice and Optimal Tax Distribution. Fuzzy Sets and Systems 53, 253–274 (1993)

[R3.1.17]   Dompere, K.K.: A Fuzzy-Decision Theory of Optimal Social Discount Rate: Collective-Choice-Theoretic. Fuzzy Sets and Systems 58, 269–301 (1993)

[R3.1.18]   Dompere, K.K.: The Theory of Social Costs and Costing for Cost- Benefit Analysis in a Fuzzy Decision Space. Fuzzy Sets and Systems 76, 1–24 (1995)

[R3.1.19]   Dompere, K.K.: The Theory of Approximate Prices: Analytical Foundations of Experimental Cost-Benefit Analysis in a Fuzzy-Decision Space. Fuzzy Sets and Systems 87, 1–26 (1997)

[R3.1.20]   Dompere, K.K.: Cost-Benefit Analysis, Benefit Accounting and Fuzzy Decisions I: Theory. Fuzzy Sets and Systems 92, 275–287 (1987)

[R3.1.21]   Dompere, K.K.: Cost-Benefit Analysis, Benefit Accounting and Fuzzy Decisions: Part II, Mental Illness in Hypothetical Community. Fuzzy Sets and Systems 100, 101–116 (1998)

[R3.1.22]   Dompere, K.K.: Cost-Benefit Analysis of Information Technology. In: Kent, A., et al. (eds.) Encyclopedia of Computer Science and Technology, vol. 41(suppl. 26), pp. 27–44. Marcal Dekker, New York (1999)

[R3.1.23]   Dompere, K.K.: Cost-Benefit Analysis and the Theory of Fuzzy Decision: Identification and Measurement Theory. STUDFUZZ, Editor-in-Chief, Prof. Janusz Kacprzyk, vol. 158. Springer, New York (2004)

[R3.1.24]   Dompere, K.K.: Cost-Benefit Analysis and the Theory of Fuzzy Decision: The Fuzzy Value Theory. STUDFUZZ, Editor-in-Chief, Prof. Janusz Kacprzyk, vol. 160. Springer, New York (2004)

[R3.1.25]   Freeman, A.: Myrick, The Measurement of Environment and Resource Values: theory and Method. Resources for the Future, Washington DC (1993)

[R3.1.26]   Gregory, R.: Interpreting Measures of Economic Loss: Evidence from Contingent Valuation and Experimental Studies. Jour. of Environmental Economics and Management 13, 325–337 (1986)

[R3.1.27]   Hanemann, W.M.: Welfare Evaluations in Contingent Valuation Experiments with Discrete Responses. American Journal of Agricultural Economics 66, 332–341 (1984)

[R3.1.28]   Harrison, G.W.: Valuing Public Goods with the Contingent Valuation Method: A Critique of Kahneman and Knetsch. Journal of Environmental Economics and Management 23, 248–257 (1992)

[R3.1.29]   Hausman, J.A. (ed.): Contingent Valuation: A Critical Assessment. North-Holland, New York (1993)

[R3.1.30]   Hoehn, J.P., Randall, A.: Too Many Proposals Pass the Benefit Cost Test. American Economic Review 79, 544–551 (1989)

[R3.1.31]   Kahneman, D.: Comments on the Contingent Valuation Method. In: Cummings, R.G., Brookshire, D.S., Schulze, W.D. (eds.) Valuing Environmental Goods: A State of the Arts Assessment of the Contingent Valuation Method, pp. 185–194. Rowman and Allanheld, Totowa (1986)

[R3.1.32]   McNeal, B.J.: On the elicitation of Preferences for alternative Therapies. New England Jour. of Medicine 306, 1259–1262 (1982)

[R3.1.33]   Portney, P.R.: The Contingent Valuation Debate: Why Economists Should Care. Jour. of Economic Perspectives 8, 3–17 (1994)

[R3.1.34]   Randall, A., et al.: Contingent Valuation Survey for Evaluating Environmental Assets. Natural Resources Jour. 23, 635–648 (1983)

[R3.1.35]   Smith, V.L.: Experiments with a Decentralized Mechanism for Public Good Decision. American Economic Review 70, 584–599 (1980)

[R3.1.36]   Smith, V.K.: An Experimental Comparison of Three Public Good Decision Mechanisms. Scandinavian Jour. of Economics 81, 198–215 (1979)

[R3.1.37]   Smith, V.K.: Indirect Revelation of demand for Public Goods: An Overview and Critique. Scot. Jour. of Political Econ. 26, 183–189 (1979)

[R3.1.38]   Tversky, A., et al.: The Framing of Decisions and the Psychology of Choice. Science 211, 453–458 (1981)

R3.2   Cost-Benefit Rationality and the Revealed Preference Approach (RPA)

[R3.2.1]   Bain, J.S.: Criteria for Undertaking Water-Resource Development. Amer. Econ. Rev. Papers and Proceedings 50(2), 310–320 (1960)

[R3.2.2]   Barsb, S.L.: Cost-Benefit Analysis and Manpower Programs. Heath & Co., Toronto (1972)

[R3.2.3]   Benefit-Cost and Policy Analysis 1973, An Aldine Annual. Aldine Pub. Co., Chicago (1974)

[R3.2.4]   Benefit-Cost Analysis and Policy Analysis 1974, An Aldine Annual. Aldine Pub. Co., Chicago (1975)

[R3.2.5]   Benefit-Cost Analysis and Policy Analysis 1971, An Aldine Annual. Aldine Pub. Co., Chicago (1972)

[R3.2.6]   Brookings Institution, Applying Benefit-Cost Analysis to Public Programs, Brookings Research Report 79, Washington, DC (1968)

[R3.2.7]   Devine, E.T.: The Treatment of Incommensurables in Cost-Benefit Analysis. Land Economics 42(3), 383–387 (1966)

[R3.2.8]   Gramlich, E.M.: Benefit-Cost Analysis of Government Programs. Prentice-Hall, Inc., Englewood Cliffs (1981)

[R3.2.9]   Haveman, R.H.: Benefit-Cost Analysis: Its Relevance to Public Investment Decisions: Comment. Quarterly, Jour. of Econ. 81(4), 695–702 (1967)

[R3.2.10]   Knesse, A.V.: Research Goals and Progress Toward them. In: Jarrett, H. (ed.) Environmental Quality in a Growing Economy, pp. 69–87. John Hopkins Press, Washington, DC (1966)

[R3.2.11]   Layard, R. (ed.): Cost-Benefit Analysis. Penguin, Baltimore (1972)

[R3.2.12]   Lesourne, J.: Cost-Benefit Analysis and Economic theory. North-Holland, New York (1975)

[R3.2.13]   Mass, A.: Benefit-Cost Analysis: Its Relevance to Public Investment Decisions. Quarterly Jour. of Economics 80(2), 208–226 (1966)

[R3.2.14]   MacDonald, J.S.: Benefits and Costs: Theoretical and Methodological Issues: Discussion. In: Somers, G.G., et al. (eds.) Proceedings of a North American Conference on Cost-Benefit Analysis of Manpower Policies, pp. 30–37. Kingston, Ontario (1969)

[R3.2.15]   Marciariello, J.A.: Dynamic Benefit-Cost Analysis. Heath and Co., Toronto (1975)

[R3.2.16]   Mishan, E.J.: Cost-Benefit Analysis. Praeger, New York (1976)

[R3.2.17]   Musgrave, R.A.: Cost-Benefit Analysis and the Theory of Public Finance. Journal of Econ. Literature 7, 797–806 (1967)

[R3.2.18]   Prest, A.R., et al.: Cost-Benefit Analysis: A Survey. In: Survey in Economic Theory, vol. III, St. Martin Press, New York (1966); Also in Economic Jour. 75, 685–705 (December 1965)

[R3.2.19]   Raynauld, A.: Benefits and Costs: Theoretical and Methodological Issue: Discussion. In: Somers, G.G., et al. (eds.) Proceedings of a North American Conference on Cost-Benefit Analysis of Manpower Policies, pp. 37–41. Kingston, Ontario (1969)

[R3.2.20]   Schwartz, H., et al. (eds.): Social and Economic Dimensions of Project Evaluation, Symposium on the Use of Socioeconomic Investment Criteria. Inter-American Dev. Bank, Washington, DC (1973)

[R3.2.21]   Solo, R.A.: Benefit-Cost Analysis and Externalities in Private Choice: Comment. Southern Econ. Jour. 34(4), 569–570 (1968)

[R3.2.22]   Somers, G.G., et al. (eds.): Cost-Benefit Analysis of Manpower Policies. Proceedings of a North American Conference. Kingston, Ontario (1969)

[R3.2.23]   United States of America Committee on Interstate and Foreign Commerce House of Representatives, Use of Cost-Benefit Analysis By Regulatory Agencies, Joint and Subcommittee on Consumer Protection and Finance, Serial 96-157 (1979)

[R3.2.24]   Water Resource Council of USA, Procedures for evaluation of national economic development (NED) benefits and costs in water resources planning (level C) final rule, Fed. Register 44, 72892-72977 (1979)

R4 Democracy, Governance and Political Markets

[R4.1]   Alexander, K.J.W.: Political Economy of Change. Blackwell, Oxford (1975)

[R4.2]   Arrow, K.: The Limits of Organization. Norton, New York (1974)

[R4.3]   Barber, J.D.: Power in Committees. Rand McNally, Chicago (1966)

[R4.4]   Beard, C.A.: The Economic Basis of Politics and Related Writings. Random House, New York (1957)

[R4.5]   Bentley, A.F.: The Process of Government. University of Chicago Press, Chicago (1907)

[R4.6]   Berle, A.A.: The Twentieth Century Capitalist Revolution. Harcourt Brace Jovanovich, New York (1954)

[R4.7]   Berle, A.A.: Power without Property: A New Development in American Political Economy. Harcourt Brace and Co., New York (1959)

[R4.8]   Bernholz, P.: Economic Policies in a Democracy. Kyklos 19, 48–80 (1966)

[R4.9]   Black, D.: The Theory of Committees and Elections. Cambridge Univ. Press, Cambridge (1958)

[R4.10]   Blaisdell Donald, C. (ed.): Unofficial Government: Pressure Groups and Lobbies, Philadelphia. The Annals, vol. 319; Sellin, T., Lambert, R.D. (Gen.eds.): The American Academy of Political and Social Science (1958)

[R4.11]   Bowles, S., Gintis, H.: Property, Community, and Contradictions of Modern Social Thought. Basic Books, New York (1986)

[R4.12]   Breton, A.: The Economic Theory of Representative Government. Aldine Pub. Co., Chicago (1974)

[R4.13]   Brenton, A., Galeotti, G., Salmon, P., Wintrobe, R. (eds.): Rational Foundations of Democratic Politics. Cambridge University Press, Cambridge (2003)

[R4.14]   Brenton, A., Galeotti, G., Salmon, P., Wintrobe, R. (eds.): Understanding Democracy. Cambridge University Press, New York (1997)

[R4.15]   Brams, S.J.: Measuring the Concentration of Power in Political Systems. American Political Science Review 62, 461–475 (1968)

[R4.16]   Breton, A.: The Economic Constitution of Federal States. University of Toronto Press, Toronto (1978)

[R4.17]   Barry, B.M.: Sociologists, Economists and Democracy. Collier-Macmillan, London (1970)

[R4.18]   Buchanan, J.M.: Individual Choice in Voting and the Market. Journal of Political Economy 62, 334–343 (1954)

[R4.19]   Buchanan, J.M.: Public Finance in Democratic Process. University of North Carolina Press, Chapel Hill (1967)

[R4.20]   Buchanan, J.M., Tullock, G.: The Calculus of Consent. University of Michigan Press, Ann Arbor (1962)

[R4.21]   Carson, R.B.: Economic Issues Today: Alternative Approaches. St. Martin's Press, New York (1991)

[R4.22]   Carson, R.B., et al. (eds.): Government in the American Economy. Heath and Co., Lexington (1973)

[R4.23]   Champlin, J.: On the Study of Power. Politics and Society 1, 91–111 (1971)

[R4.24]   Clark, J.M.: Alternative to Selfdom. Random House/Vintage Books, New York (1960)

[R4.25]   Clark, J.M.: Social Control of Business. McGraw-Hill, New York (1939)

[R4.26]   Cohen, C. (ed.): Communism, Fascism and Democracy: The Theoretical Foundations I&II. Random House, New York (1962)

[R4.27]   Cornforth, M.: The Open Philosophy and the Open Society. International Publishers, New York (1968)

[R4.28]   Crossman, R.H.S.: The Politics of Socialism. Athenaeum Pub., New York (1965)

[R4.29]   Dahl, R.A.: The Concept of Power. Behavioral Science 2, 201–215 (1957)

[R4.30]   Dahl, R.A.: Who Governs? Democracy and Power in an American City. Yale University Press, New Haven (1961)

[R4.31]   Dell, E.: Political Responsibility and Industry. Allen & Unwin, London (1973)

[R4.32]   Diermier, D., Merlo, A.: Government Turnover in Parliamentary Democracies. Journal of Economic Theory 94, 46–79 (2000)

[R4.33]   Dobb, M.: Studies in the Development of Capitalism. International Publishers, New York (1970)

[R4.34]   Domhoff, W.: Who Rules America? Prentice-Hall, Englewood Cliffs (1967)

[R4.35]   Downs, A.: An Economic Theory of Democracy. Harper and Row, New York (1957)

[R4.36]   Downs, A.: Inside Bureaucracy. Little Brown, Boston (1966)

[R4.37]   Easton, D. (ed.): Varieties of Political Theory. Prentice-Hall, Englewood Cliffs (1968)

[R4.38]   Farquharson, R.: Theory of Voting. Yale University Press, New Haven (1969)

[R4.39]   Fiorina, M.P.: Majority Rule Models and Legislative Elections. Journal of Politics 41, 1081–1104 (1979)

[R4.40]   Fishkin, J.S.: Democracy and Deliberation: New Direction for Democratic Reform. Yale University Press, New Haven (1991)

[R4.41]   Friedman, M.: Capitalism and Freedom. University of Chicago Press, Chicago (1962)

[R4.42]   Frisch, H. (ed.): Schumpeterian Economics. Praeger, New York (1982)

[R4.43]   Fulkerson, D.R.: Networks, Frames, Blocking Systems. In: Mathematics of Decision Sciences, vol. 1. American Mathematical Society, Providence (1968)

[R4.44]   Galbraith, J.K.: The Affluent Society. Houghton Mifflin, Boston (1971)

[R4.45]   Galbraith, J.K.: Economics in Perspective. Houghton Mifflin, Boston (1987)

[R4.46]   Galbraith, J.K.: The New Industrial State. Houghton Mifflin, Boston (1967)

[R4.47]   Galbraith, J.K.: Economics and Public Purpose. Houghton Mifflin, Boston (1967)

[R4.48]   Galbraith, J.K.: American Capitalism: The Concept of Countervailing Powers. Houghton Mifflin, Boston (1956)

[R4.49]   Gamson, W.A.: Coalition Formation at Presidential Nominating Conventions. American Journal of Sociology 68, 157–171 (1962)

[R4.50]   Gamson, W.A.: The Theory of Coalition Formation. American Sociological Review 26, 373–382 (1961)

[R4.51]   Gibbard, A.: Manipulation of Voting Schemes: A General Result. Econometrica 41, 587–601 (1973)

[R4.52]   Ginsberg, B.: Elections and Public Policy. American Political Science Review 68, 41–49 (1976)

[R4.53]   Graham, F.D.: Social Goals and Economic Institutions. Princeton University Press, Princeton (1942)

[R4.54]   Hahn, E.L.: Revival of Political Economy: The Wrong Issues and the Wrong Argument. Economic Record 51, 360–364 (1975)

[R4.55]   Harris, S.E.: The Economics of the Two Political Parties. Macmillan, New York (1962)

[R4.56]   Hayek, F.: The Road to Serfdom. University of Chicago Press, Chicago (1944)

[R4.57]   Heertje, A.: Schumpeter's Vision: Capitalism Socialism and Democracy after 40 Years. Praeger, New York (1981)

[R4.58]   Heilbroner, R.: The Nature and Logic of Capitalism. Norton, New York (1985)

[R4.59]   Heller, W.W.: New Dimensions of Political Economy. Harvard University Press, Cambridge (1966)

[R4.60]   Hibbs Jr., D.: Political Parties and Macroeconomic Policy. American Political Science Review 71, 1467–1487 (1977)

[R4.61]   Ingberman, D.E.: Running Against the Status Quo: Institutions for Direct Democracy Referenda and Allocation over Time. Public Choice 46, 19–43 (1985)

[R4.62]   Key Jr., V.O.: Politics, Parties, and Pressure Groups. Thomas Y. Cromwell, New York (1964)

[R4.63]   Kirk, R.: The Conservative Mind. Chicago, Regnery (1954)

[R4.64]   Kitschelt, H.: Linkages between Citizens and Politicians in Democratic Politics. Comparative Political Studies 33(6/7), 845–879

[R4.65]   Kitschelt, H., Wilkinson, S. (eds.): Citizen-Politician Linkages in Democratic Politics. Cambridge University Press, Cambridge (2006)

[R4.66]   Knight, F.: Freedom and Reform. Harper & Row, New York (1947)

[R4.67]   Knoke, D.: Change and Continuity in American Politics: The Social Basis of Political Parties. The Johns Hopkins University Press, Baltmore (1976)

[R4.68]   Kollman, K., Miller, J.H., Page, S.E. (eds.): Computational Models in Political Economy. MIT Press, Cambridge (2003)

[R4.69]   La Palombra, J.: Bureaucracy and Political Development. Princeton University Press, Princeton (1963)

[R4.70]   Lindbeck, A.: The Political Economy of the New Left. Harper and Row, New York (1977)

[R4.71]   Lohmann, S.: An Information Rationale for the Power of Special Interest. American Political Science Review 92(4), 809–827 (1998)

[R4.72]   Malthus, T.R.: Definitions in Political Economy. Augustus M. Kelley, New York (1963)

[R4.73]   March, J.G.: An Introduction to the Theory and Measurement of Influence. American Political Science Review 49, 431–451 (1955)

[R4.74]   March, J.G.: The Power of Power. In: Easton, D. (ed.) Varieties of Political Theory, pp. 39–70. Prentice-Hall, Englewood Cliffs (1968)

[R4.75]   Martindale, D. (ed.): National Character in the Perspective of the Social Sciences, Philadelphia. The Annals, vol. 370. The American Academy of Political and Social Science (1967); Sellin, T., Lambert, R.D. (Gen. eds.)

[R4.76]   Marx, K.: Contribution to the Critique of Political Economy. Charles H. Kerr and Co., Chicago (1904)

[R4.77]   Marx, K.: Capital, vol. 1, 2, 3. Progress Publishers, Moscow (1887)

[R4.78]   Marx, K.: Theories of Surplus –Value, vol. Part 1, 1963, Part II, 1968, Part III. Progress Publishers, Moscow (1971)

[R4.79]   Maschler, M.: The Power of Coalition. Management Science 10, 8–29 (1963)

[R4.80]   Mayhew, D.: Congress: The Electoral Connection. Yale University Press, New Haven (1974)

[R4.81]   Midgaard, K.: Strategy and Ethics in international Politics. Cooperation and Conflict, 4, 224–240 (1770)

[R4.82]   Mueller, D.C.: Constitutional Democracy and Social Welfare. Quart. Jour. of Econ. 87, 60–80 (1973)

[R4.83]   Niskanen, W.A.: Bureaucracy and Representative Government. Aldine, Chicago (1971)

[R4.84]   Noll, R., Fiorina, M.: Voters, Bureaucrates and Legislators. Journal of Public Economics 7(32), 239–254 (1978)

[R4.85]   Normanton, E.L.: The Accountability and Audit of Governments. Praeger, New York (1966)

[R4.86]   Phelps, E.S. (ed.): Private Wants and Public Needs. W. W. Norton, New York (1957)

[R4.87]   Piattoni, S. (ed.): Clientelism, Interests, and Democratic Representation. Cambridge University Press, Cambridge (2001)

[R4.88]   Popkin, S.L.: The Reasoning Voter: Communication and Persuasion in Presidential Campaigns. University of Chicago Press, Chicago (1991)

[R4.89]   Preston, N.S.: Politics, Economics and Power: Ideology and Practice Under Capitalism, Socialism, Communism and Fascism. Macmillan, London (1967)

[R4.90]   Richter, M.K.: Coalitions, Core, and Competition. Journal of Economic Theory 3, 323–334 (1971)

[R4.91]   Riker, W.H.: A Test of the Adequacy of the Power Index. Behavioral Science 4, 120–131 (1959)

[R4.92]   Riker, W.H.: The Theory of Political Coalitions. Yale University Press, New Haven (1962)

[R4.93]   Riker, W.H.: Some Ambiguities in the Notion of Power. American Political Science Review 58, 341–349 (1964)

[R4.94]   Robbins, L.O.: Political Economy Past and Present. Macmillan, London (1976)

[R4.95]   Rostow, E.: Planning for Freedom: The Public Law of American Capitalism. Yale University Press, New Haven (1959)
[R4.96]   Rowley, C.K. (ed.): Democracy and Public Choice. Basil Blackwell, New York (1987)
[R4.97]   Salamon, L.M., Siegfried, J.J.: Economic Power and Political Influence: The impact of Industry Structure on Public Policy. American Political Science Review 71, 1026–1043 (1977)
[R4.98]   Schotter, A.: The Economic Theory of Social Institutions. Cambridge University Press, Cambridge (1981)
[R4.99]   Schumpter, J.A.: The Theory of Economic Development. Harvard University Press, Cambridge (1934)
[R4.100]   Schumpter, J.A.: Capitalism, Socialism and Democracy. Harper & Row, New York (1950)
[R4.101]   Schumpeter, J.A.: March to Socialism. American Economic Review 40, 446–456 (1950)
[R4.102]   Schumpeter, J.A.: Theoretical Problems of Economic Growth. Journal of Economic History 8, 1–9 (1947)
[R4.103]   Schumpeter, J.A.: The Analysis of Economic Change. Review of Economic Statistics 17, 2–10 (1935)
[R4.104]   Shapley, L.S.: Pure Competition, Coalitional Power, and Fair Division. International Economic Review 10, 337–362 (1969)
[R4.105]   Shepsle, K., Weingast, B.: When Do Rules of Procedure Matter? The Journal of Politics 46, 206–221
[R4.106]   Sigel, R. (ed.): Political Socialization: Its Role in the Political Process Philadelphia. The Annals, vol. 361. The American Academy of Political and Social Science (1965); Sellin, T., Lambert, R.D. (Gen.eds.)
[R4.107]   Stigler, G.J.: General Economic Conditions and National Elections. American Economic Review 63, 160–167 (1973)
[R4.108]   Stokes, D.E., Campbell, A., Miller, W.E.: Components of Electoral Decision. American Political Science Review 52, 367–387 (1958)
[R4.109]   Thurow, L.C.: Dangerous Currents. Random House, New York (1983)
[R4.110]   Tufte, E.R.: The Relationship between Seats and Votes in Two-Party System. American Political Science Review 67, 540–554 (1973)
[R4.111]   Tullock, G.: The Politics of Bureaucracy. Public Affairs Press, Washington, DC (1965)
[R4.112]   Tullock, G.: Bureaucracy: The Selected Works Vol, vol. 6. Liberty Fund, Indianapolis (2005)
[R4.113]   Tullock, G.: The Economics of Politics: The Selected Works, vol. 4. Liberty Fund, Indianapolis (2004)
[R4.114]   Tullock, G.: Virginia Political Economy: The Selected Works, vol. 1. Liberty Fund, Indianapolis (2005)
[R4.115]   Tullock, G.: The Calculus of Consent: Logical Foundations of Constitutional Democracy, The Selected Works, vol. 2. Liberty Fund, Indianapolis (2005)
[R4.116]   Tullock, G.: Some Problems of Majority Voting. Journal of Political Economy 67, 571–579 (1959)
[R4.117]   Windmuller, J.P. (ed.): Industrial Democracy in International Perspective, Philadelphia, The Annals, vol. 431. The American Academy of Political and Social Science (1977); Lambert, R.D., Heston, A.W. (Gen. eds.)
[R4.118]   Winters, T.: Party Control and Policy Change. American Journal of Political Science 20, 597–636 (1976)

[R4.119]   Yergin, D., Stanislaw, J.: The Commanding Heights: The Battle Between Government and the Marketplace that is Remaking the Modern Word. Simon & Schuster, New York (1999)

[R4.120]   Young, H.P.: Power, Prices and Income in Voting Systems. Mathematical Programming 14, 129–148 (1978)

[R4.121]   Young, H.P.: The Allocation of Funds in Lobbying and Campaigning. Behavioral Science 23, 21–31 (1978)

R5   Fuzzy Game Theory

[R5.1]   Aubin, J.P.: Cooperative Fuzzy Games. Mathematics of Operations Research 6, 1–13 (1981)

[R5.2]   Aubin, J.P.: Mathematical Methods of Game and Economics Theory. North Holland, New York (1979)

[R5.3]   Butnaria, D.: Fuzzy Games: A description of the concepts. Fuzzy Sets and Systems 1, 181–192 (1978)

[R5.4]   Butnaria, D.: Stability and shapely value for a n–persons Fuzzy Games. Fuzzy Sets and Systems 4(1), 63–72 (1980)

[R5.5]   Nurmi, H.: A Fuzzy Solution to a Majority Voting Game. Fuzzy Sets and Systems 5, 187–198 (1981)

[R5.6]   Regade, R.K.: Fuzzy Games in the Analysis of Options. Jour. of Cybernetics 6, 213–221 (1976)

[R5.7]   Spillman, B., et al.: Coalition Analysis with Fuzzy Sets. Kybernetes 8, 203–211 (1979)

[R5.8]   Wernerfelt, B.: Semifuzzy Games. Fuzzy Sets and systems 19, 21–28 (1986)

R6   Fuzzy Mathematics and Optimal Rationality

[R6.1]   Bandler, W., et al.: Fuzzy Power Sets and Fuzzy Implication Operators. Fuzzy Sets and Systems 4(1), 13–30 (1980)

[R6.2]   Banon, G.: Distinction between Several Subsets of Fuzzy Measures. Fuzzy Sets and Systems 5(3), 291–305 (1981)

[R6.3]   Bellman, R.E.: Mathematics and Human Sciences. In: Wilkinson, J., et al. (eds.) The Dynamic Programming of Human Systems, pp. 11–18. MSS Information Corp., New York (1973)

[R6.4]   Bellman, R.E., Glertz, M.: On the Analytic Formalism of the Theory of Fuzzy Sets. Information Science 5, 149–156 (1973)

[R6.5]   Brown, J.G.: A Note On Fuzzy Sets. Information and Control 18, 32–39 (1971)

[R6.6]   Butnariu, D.: Fixed Points For Fuzzy Mapping. Fuzzy Sets and Systems 7(2), 191–207 (1982)

[R6.7]   Butnariu, D.: Decompositions and Range For Additive Fuzzy Measures. Fuzzy Sets and Systems 10(2), 135–155 (1983)

[R6.8]   Cerruti, U.: Graphs and Fuzzy Graphs. In: Fuzzy Information and Decision Processes, pp. 123–131. North-Holland, New York (1982)

[R6.9]   Chakraborty, M.K., et al.: Studies in Fuzzy Relations Over Fuzzy Subsets. Fuzzy Sets and Systems 9(1), 79–89 (1983)

[R6.10]   Chang, C.L.: Fuzzy Topological Spaces. J. Math. Anal. and Applications 24, 182–190 (1968)

[R6.11] Chang, S.S.L.: Fuzzy Mathematics, Man and His Environment. IEEE Transactions on Systems, Man and Cybernetics SMC-2, 92–93 (1972)

[R6.12] Chang, S.S.L., et al.: On Fuzzy Mathematics and Control. IEEE Transactions, System, Man and Cybernetics SMC-2, 30–34 (1972)

[R6.13] Chang, S.S.: Fixed Point Theorems for Fuzzy Mappings. Fuzzy Sets and Systems 17, 181–187 (1985)

[R6.14] Chapin, E.W.: An Axiomatization of the Set Theory of Zadeh. Notices, American Math. Society 687-02-4 754 (1971)

[R6.15] Chaudhury, A.K., Das, P.: Some Results on Fuzzy Topology on Fuzzy Sets. Fuzzy Sets and Systems 56, 331–336 (1993)

[R6.16] Cheng-Zhong, L.: Generalized Inverses of Fuzzy Matrix. In: Gupta, M.M., et al. (eds.) Approximate Reasoning in Decision Analysis, pp. 57–60. North Holland, New York (1982)

[R6.17] Chitra, H., Subrahmanyam, P.V.: Fuzzy Sets and Fixed Points. Jour. of Mathematical Analysis and Application 124, 584–590 (1987)

[R6.18] Cohn, D.L.: Measure Theory. Birkhauser, Boston (1980)

[R6.19] Cohen, P.J., Hirsch, R.: Non-Cantorian Set Theory. Scientific America, 101–116 (December 1967)

[R6.20] Czogala, J., et al.: Fuzzy Relation Equations on a Finite Set. Fuzzy Sets and Systems 7(1), 89–101 (1982)

[R6.21] Das, P.: Fuzzy Topology on Fuzzy Sets: Product Fuzzy Topology and Fuzzy Topological Groups. Fuzzy Sets and Systems 100, 367–372 (1998)

[R6.22] DiNola, A., et al. (eds.): The Mathematics of Fuzzy Systems. Verlag TUV Rheinland, Koln (1986)

[R6.23] DiNola, A., et al.: On Some Chains of Fuzzy Sets. Fuzzy Sets and Systems 4(2), 185–191 (1980)

[R6.24] Dombi, J.: A General Class of Fuzzy Operators, the DeMorgan Class of Fuzzy Operators and Fuzzy Measures Induced by Fuzzy Operators. Fuzzy Sets and Systems 8(2), 149–163 (1982)

[R6.25] Dubois, D., Prade, H.: Towards Fuzzy Differential Calculus, Part I: Integration of Fuzzy Mappings. Fuzzy Sets and Systems 8(1), 1–17 (1982)

[R6.26] Dubois, D., Prade, H.: Towards Fuzzy Differential Calculus, Part 2: Integration on Fuzzy Intervals. Fuzzy Sets and Systems 8(2), 105–116 (1982)

[R6.27] Dubois, D., Prade, H.: Towards Fuzzy Differential Calculus, Part 3: Differentiation. Fuzzy Sets and Systems 8(3), 225–233 (1982)

[R6.28] Dubois, D., Prade, H.: Fuzzy Sets and Systems. Academic Press, New York (1980)

[R6.29] Dubois, D.: Fuzzy Real Algebra: Some Results. Fuzzy Sets and Systems 2(4), 327–348 (1979)

[R6.30] Dubois, D., Prade, H.: Gradual Inference rules in approximate reasoning. Information Sciences 61(1-2), 103–122 (1992)

[R6.31] Dubois, D., Prade, H.: On the combination of evidence in various mathematical frameworks. In: Flamm, J., Luisi, T. (eds.) Reliability Data Collection and Analysis, pp. 213–241. Kluwer, Boston (1992)

[R6.32] Dubois, D., Prade, H. (eds.): Readings in Fuzzy Sets for Intelligent Systems. Morgan Kaufmann, San Mateo (1993)

[R6.33] Dubois, D., Prade, H.: Fuzzy sets and probability: Misunderstanding, bridges and gaps. In: Proc. Second IEEE Intern. Conf. on Fuzzy Systems, San Francisco, pp. 1059–1068 (1993)

[R6.34]   Dubois, D., Prade, H.: A survey of belief revision and updating rules in various uncertainty models. Intern. J. of Intelligent Systems 9(1), 61–100 (1994)

[R6.35]   Erceg, M.A.: Functions, Equivalence Relations, Quotient Spaces and Subsets in Fuzzy Set Theory. Fuzzy Sets and Systems 3(1), 79–92 (1980)

[R6.36]   Feng, Y.-J.: A Method Using Fuzzy Mathematics to Solve the Vectormaximum Problem. Fuzzy Sets and Systems 9(2), 129–136 (1983)

[R6.37]   Filev, D.P., et al.: A Generalized Defuzzification Method via Bag Distributions. Intern. Jour. of Intelligent Systems 6(7), 687–697 (1991)

[R6.38]   Foster, D.H.: Fuzzy Topological Groups. Journal of Math. Analysis and Applications 67, 549–564 (1979)

[R6.39]   Goetschel Jr., R., et al.: Topological Properties of Fuzzy Number. Fuzzy Sets and Systems 10(1), 87–99 (1983)

[R6.40]   Goguen, J.A.: Mathematical Representation of Hierarchically Organized System. In: Attinger, E.O. (ed.) Global System Dynamics, pp. 111–129. S. Karger, Berlin (1970)

[R6.41]   Goodman, I.R.: Fuzzy Sets As Random Level Sets: Implications and Extensions of the Basic Results. In: Lasker, G.E. (ed.) Applied Systems and Cybernetics. Fuzzy Sets and Systems, vol. VI, pp. 2756–2766. Pergamum Press, New York (1981)

[R6.42]   Goodman, I.R.: Fuzzy Sets As Equivalence Classes of Random Sets. In: Yager, R.R. (ed.) Fuzzy Set and Possibility Theory: Recent Development, pp. 327–343. Pergamon Press, New York (1992)

[R6.43]   Gupta, M.M., et al. (eds.): Fuzzy Antomata and Decision Processes. North-Holland, New York (1977)

[R6.44]   Gupta, M.M., Sanchez, E. (eds.): Fuzzy Information and Decision Processes. North-Holland, New York (1982)

[R6.45]   Higashi, M., Klir, G.J.: On measure of fuzziness and fuzzy complements. Intern. J. of General Systems 8(3), 169–180 (1982)

[R6.46]   Higashi, M., Klir, G.J.: Measures of uncertainty and information based on possibility distributions. International Journal of General Systems 9(1), 43–58 (1983)

[R6.47]   Higashi, M., Klir, G.J.: On the notion of distance representing information closeness: Possibility and probability distributions. Intern. J. of General Systems 9(2), 103–115 (1983)

[R6.48]   Higashi, M., Klir, G.J.: Resolution of finite fuzzy relation equations. Fuzzy Sets and Systems 13(1), 65–82 (1984)

[R6.49]   Higashi, M., Klir, G.J.: Identification of fuzzy relation systems. IEEE Trans. on Systems, Man, and Cybernetics 14(2), 349–355 (1984)

[R6.50]   Ulrich, H.: A Mathematical Theory of Uncertainty. In: Yager, R.R. (ed.) Fuzzy Set and Possibility Theory: Recent Developments, pp. 344–355. Pergamon, New York (1982)

[R6.51]   Jin-wen, Z.: A Unified Treatment of Fuzzy Set Theory and Boolean Valued Set theory: Fuzzy Set Structures and Normal Fuzzy Set Structures. Jour. Math. Anal. and Applications 76(1), 197–301 (1980)

[R6.52]   Kandel, A.: Fuzzy Mathematical Techniques with Applications. Addison-Wesley, Reading (1986)

[R6.53]   Kandel, A., Byatt, W.J.: Fuzzy Processes. Fuzzy Sets and Systems 4(2), 117–152 (1980)

[R6.54]   Kaufmann, A., Gupta, M.M.: Introduction to fuzzy arithmetic: Theory and applications. Van Nostrand Rheinhold, New York (1991)

[R6.55]   Kaufmann, A.: Introduction to the Theory of Fuzzy Subsets, vol. 1.
           Academic Press, New York (1975)
[R6.56]   Kaufmann, A.: Theory of Fuzzy Sets. Merson Press, Paris (1972)
[R6.57]   Kaufmann, A., et al.: Fuzzy Mathematical Models in Engineering and
           Management Science. North-Holland, New York (1988)
[R6.58]   Kim, K.H., et al.: Generalized Fuzzy Matrices. Fuzzy Sets and
           Systems 4(3), 293–315 (1980)
[R6.59]   Klement, E.P.: Fuzzy σ – Algebras and Fuzzy Measurable Functions. Fuzzy
           Sets and Systems 4, 83–93 (1980)
[R6.60]   Klement, E.P.: Characterization of Finite Fuzzy Measures Using Markoff-
           kernels. Journal of Math. Analysis and Applications 75, 330–339 (1980)
[R6.61]   Klement, E.P.: Construction of Fuzzy σ – Algebras Using Triangular
           Norms. Journal of Math. Analysis and Applications 85, 543–565 (1982)
[R6.62]   Klement, E.P., Schwyhla, W.: Correspondence Between Fuzzy Measures
           and Classical Measures. Fuzzy Sets and Systems 7(1), 57–70 (1982)
[R6.63]   Klir, G., Yuan, B.: Fuzzy Sets and Fuzzy Logic. Prentice Hall, Upper
           Saddle River (1995)
[R6.64]   Kokawa, M., et al.: Fuzzy-Theoretical Dimensionality Reduction Method of
           Multi-Dimensional Quality. In: Gupta, M.M., Sanchez, E. (eds.) Fuzzy
           Information and Decision Processes, pp. 235–250. North-Holland, New
           York (1982)
[R6.65]   Kramosil, I., et al.: Fuzzy Metrics and Statistical Metric Spaces.
           Kybernetika 11, 336–344 (1975)
[R6.66]   Kruse, R.: On the Construction of Fuzzy Measures. Fuzzy Sets and
           Systems 8(3), 323–327 (1982)
[R6.67]   Kruse, R., et al.: Foundations of Fuzzy Systems. John Wiley and Sons, New
           York (1994)
[R6.68]   Lasker, G.E. (ed.): Applied Systems and Cybernetics. Fuzzy Sets and
           Systems, vol. VI. Pergamon Press, New York (1981)
[R6.69]   Lake, L.: Fuzzy Sets, Multisets and Functions I. London Math. Soc. 12(2),
           323–326 (1976)
[R6.70]   Lientz, B.P.: On Time Dependent Fuzzy Sets. Inform. Science 4, 367–376
           (1972)
[R6.71]   Lowen, R.: On the Existence of Natural Non-Topological Fuzzy
           Topological Space. Haldermann Verlag, Berlin (1986)
[R6.72]   Martin, H.W.: Weakly Induced Fuzzy Topological Spaces. Jour. Math.
           Anal. and Application 78, 634–639 (1980)
[R6.73]   Michalek, J.: Fuzzy Topologies. Kybernetika 11, 345–354 (1975)
[R6.74]   Mizumoto, M., Tanaka, K.: Some Properties of Fuzzy Numbers. In: Gupta,
           M.M., et al. (eds.) Advances in Fuzzy Sets Theory and Applications, pp.
           153–164. North Holland, Amsterdam (1979)
[R6.75]   Negoita, C.V., et al.: Applications of Fuzzy Sets to Systems Analysis.
           Wiley and Sons, New York (1975)
[R6.76]   Negoita, C.V.: Representation Theorems For Fuzzy Concepts.
           Kybernetes 4, 169–174 (1975)
[R6.77]   Negoita, C.V., et al.: On the State Equation of Fuzzy Systems.
           Kybernetes 4, 213–214 (1975)
[R6.78]   Negoita, C.V.: Fuzzy Sets in Topoi. Fuzzy Sets and Systems 8(1), 93–99
           (1982)
[R6.79]   Netto, A.B.: Fuzzy Classes. Notices, American Mathematical Society 68T-
           H28, 945 (1968)

[R6.80]   Nguyen, H.T.: Possibility Measures and Related Topics. In: Gupta, M.M., et al. (eds.) Approximate Reasoning in Decision Analysis, pp. 197–202. North Holland, New York (1982)

[R6.81]   Nowakowska, M.: Some Problems in the Foundations of Fuzzy Set Theory. In: Gupta, M.M., et al. (eds.) Approximate Reasoning in Decision Analysis, pp. 349–360. North Holland, New York (1982)

[R6.82]   Ovchinnikov, S.V.: Structure of Fuzzy Binary Relations. Fuzzy Sets and Systems 6(2), 169–195 (1981)

[R6.83]   Pedrycz, W.: Fuzzy Relational Equations with Generalized Connectives and Their Applications. Fuzzy Sets and Systems 10(2), 185–201 (1983)

[R6.84]   Raha, S., et al.: Analogy Between Approximate Reasoning and the Method of Interpolation. Fuzzy Sets and Systems 51(3), 259–266 (1992)

[R6.85]   Ralescu, D.: Toward a General Theory of Fuzzy Variables. Jour. of Math. Analysis and Applications 86(1), 176–193 (1982)

[R6.86]   Rao, M.B., et al.: Some Comments On Fuzzy Variables. Fuzzy Sets and Systems 6(2), 285–292 (1981)

[R6.87]   Rodabaugh, S.E.: Fuzzy Arithmetic and Fuzzy Topology. In: Lasker, G.E. (ed.) Applied Systems and Cybernetics. Fuzzy Sets and Systems, vol. VI, pp. 2803–2807. Pergamon Press, New York (1981)

[R6.88]   Rodabaugh, S., et al. (eds.): Application of Category Theory to Fuzzy Subsets. Kluwer, Boston (1992)

[R6.89]   Roubens, M., et al.: Linear Fuzzy Graphs. Fuzzy Sets and Systems 10(1), 79–86 (1983)

[R6.90]   Rosenfeld, A.: Fuzzy Groups. Jour. Math. Anal. Appln. 35, 512–517 (1971)

[R6.91]   Rosenfeld, A.: Fuzzy Graphs. In: Zadeh, L.A., et al. (eds.) Fuzzy Sets and Their Applications to Cognitive and Decision Processes, pp. 77–95. Academic Press, New York (1974)

[R6.92]   Rubin, P.A.: A Note on the Geometry of Reciprocal Fuzzy Relations. Fuzzy Sets and Systems 7(3), 307–309 (1982)

[R6.93]   Ruspini, E.H.: Recent Developments in Mathematical Classification Using Fuzzy Sets. In: Lasker, G.E. (ed.) Applied Systems and Cybernetics. Fuzzy Sets and Systems, vol. VI, pp. 2785–2790. Pergamon Press, New York (1981)

[R6.94]   Sanchez, E.: Resolution of Composite Fuzzy Relation Equations. Information and Control 3, 39–47 (1976)

[R6.95]   Santos, E.S.: Maximin, Minimax and Composite Sequential Machines. Jour. Math. Anal. and Appln. 24, 246–259 (1968)

[R6.96]   Santos, E.S.: Maximin Sequential Chains. Jour. of Math. Anal. and Appln. 26, 28–38 (1969)

[R6.97]   Santos, E.S.: Fuzzy Algorithms. Inform. and Control 17, 326–339 (1970)

[R6.98]   Sarkar, M.: On Fuzzy Topological Spaces. Jour. Math. Anal. Appln. 79, 384–394 (1981)

[R6.99]   Skala, H., et al. (eds.): Aspects of Vagueness. D. Reidel, Boston (1984)

[R6.100]   Slowinski, R., Teghem, J. (eds.): Stochastic versus Fuzzy Approaches to Multiobjective Mathematical Programming Under Uncertainty. Kluwer, Dordrecht (1990)

[R6.101]   Stein, N.E., Talaki, K.: Convex Fuzzy Random Variables. Fuzzy Sets and Systems 6(3), 271–284 (1981)

[R6.102]   Sugeno, M.: Inverse Operation of Fuzzy Integrals and Conditional Fuzzy Measures. Transactions SICE 11, 709–714 (1975)

[R6.103]   Taylor, J.G. (ed.): Mathematical Approaches to Neural Networks. North-Holland, New York

[R6.104]   Wright, C.: On the Coherence of Vague Predicates. Synthese 3, 325–365
             (1975)
[R6.105]   Yager, R.R., Filver, D.P.: Essentials of Fuzzy Modeling and Control. John
             Wiley and Sons, New York (1994)
[R6.106]   Perano, T., et al.: Fuzzy Systems Theory and its Applications. Academic
             Press, New York (1992)
[R6.107]   Terano, T., et al.: Applied Fuzzy Systems. AP Professional, New York
             (1994)
[R6.108]   Triantaphyllon, E., et al.: The Problem of Determining Membership Values
             in Fuzzy Sets in Real World Situations. In: Brown, D.E., et al. (eds.)
             Operations Research and Artificial Intelligence: The Integration of
             Problem-solving Strategies, pp. 197–214. Kluwer, Boston (1990)
[R6.109]   Tsichritzis, D.: Participation Measures. Jour. Math. Anal. and Appln. 36,
             60–72 (1971)
[R6.110]   Tsichritzis, D.: Approximation and Complexity of Functions on the
             Integers. Inform. Science 4, 70–86 (1971)
[R6.111]   Turksens, I.B.: Four Methods of Approximate Reasoning with Interval-
             Valued Fuzzy Sets. Intern. Journ. of Approximate Reasoning 3(2), 121–142
             (1989)
[R6.112]   Turksen, I.B.: Measurement of Membership Functions and Their
             Acquisition. Fuzzy Sets and Systems 40(1), 5–38 (1991)
[R6.113]   Verdegay, J., et al.: The Interface Between Artificial Intelligence and
             Operations Research in Fuzzy Environment. Verlag TUV Rheinland, Koln
             (1989)
[R6.114]   Wang, L.X.: Adaptive Fuzzy Sets and Control: Design and Stability
             Analysis. Prentice Hall, Englewood Cliffs (1994)
[R6.115]   Wang, P.P. (ed.): Advances in Fuzzy Sets, Possibility Theory, and
             Applications. Plenum Press, New York (1983)
[R6.116]   Wang, P.P. (ed.): Advances in Fuzzy Theory and Technology, vol. 1.
             Bookwright Press, Durham (1992)
[R6.117]   Wang, Z., Klir, G.: Fuzzy Measure Theory. Plenum Press, New York
             (1992)
[R6.118]   Wang, P.Z., et al. (eds.): Between Mind and Computer: Fuzzy Science and
             Engineering. World Scientific Press, Singapore (1993)
[R6.119]   Wang, P.Z.: Contactability and Fuzzy Variables. Fuzzy Sets and
             Systems 8(1), 81–92 (1982)
[R6.120]   Wang, S.: Generating Fuzzy Membership Functions: A Monotonic Neural
             Network Model. Fuzzy Sets and Systems 61(1), 71–82 (1994)
[R6.121]   Whalen, T., et al.: Usuality, Regularity, and Fuzzy Set Logic. Intern. Jour.
             of Approximate Reasoning 6(4), 481–504 (1992)
[R6.122]   Wierzchon, S.T.: An Algorithm for Identification of Fuzzy Measure. Fuzzy
             Sets and Systems 9(1), 69–78 (1983)
[R6.123]   Wong, C.K.: Fuzzy Topology: Product and Quotient Theorems. Journal of
             Math. Analysis and Applications 45, 512–521 (1974)
[R6.124]   Wong, C.K.: Fuzzy Points and Local Properties of Fuzzy Topology. Jour.
             Math. Anal. and Appln. 46, 316–328 (1987)
[R6.125]   Wong, C.K.: Fuzzy Topology. In: Zadeh, A., et al. (eds.) Fuzzy Sets and
             Their Applications to Cognitive and Decision Processes, pp. 171–190.
             Academic Press, New York (1974)
[R6.126]   Wong, C.K.: Categories of Fuzzy Sets and Fuzzy Topological Spaces. Jour.
             Math. Anal. and Appln. 53, 704–714 (1976)

[R6.127]   Wygralak, M.: Fuzzy Inclusion and Fuzzy Equality of two Fuzzy Subsets. Fuzzy Sets and Systems 10(2), 157–168 (1983)

[R6.128]   Yager, R.R.: On the Lack of Inverses in Fuzzy Arithmetic. Fuzzy Sets and Systems 4(1), 73–82 (1980)

[R6.129]   Yager, R.R. (ed.): Fuzzy Set and Possibility Theory: Recent Development. Pergamon Press, New York (1992)

[R6.130]   Yager, R.R.: Fuzzy Subsets with Uncertain Membership Grades. IEEE Transactions on Systems, Man and Cybernetics 14(2), 271–275 (1984)

[R6.131]   Yager, R.R., et al.: Essentials of Fuzzy Modeling and Control. John Wiley, New York (1994)

[R6.132]   Yager, R.R., et al. (eds.): Fuzzy Sets, Neural Networks, and Soft Computing. Nostrand Reinhold, New York (1994)

[R6.133]   Yager, R.R.: On the Theory of Fuzzy Bags. Intern. Jour. of General Systems 13(1), 23–37 (1986)

[R6.134]   Yager, R.R.: Cardinality of Fuzzy Sets via Bags. Mathematical Modelling 9(6), 441–446 (1987)

[R6.135]   Zadeh, L.A.: A Computational Theory of Decompositions. Intern. Jour. of Intelligent Systems 2(1), 39–63 (1987)

[R6.136]   Zadeh, L.A.: The Birth and Evolution of Fuzzy Logic. Intern. Jour. of General Systems 17(2-3), 95–105 (1990)

[R6.137]   Zadeh, L.A., et al.: Fuzzy Logic for the Management of Uncertainty. John Wiley, New York (1992)

[R6.138]   Zimmerman, H.J.: Fuzzy Set Theory and Its Applications. Kluwer, Boston (1985)

R7 Fuzzy Optimization and Decision-Choice Rationality

[R7.1]   Bose, R.K., Sahani, D.: Fuzzy Mappings and Fixed Point Theorems. Fuzzy Sets and Systems 21, 53–58 (1987)

[R7.2]   Buckley, J.J.: Fuzzy Programming And the Pareto Optimal Set. Fuzzy Set and Systems 10(1), 57–63 (1983)

[R7.3]   Butnariu, D.: Fixed Points for Fuzzy Mappings. Fuzzy Sets and Systems 7, 191–207 (1982)

[R7.4]   Carlsson, G.: Solving Ill-Structured Problems Through Well Structured Fuzzy Programming. In: Brans, J.P. (ed.) Operation Research 81, pp. 467–477. North-Holland, Amsterdam (1981)

[R7.5]   Carlsson, C.: Tackling an AMDM - Problem with the Help of Some Results From Fuzzy Set Theory. European Journal of Operational Research 10(3), 270–281 (1982)

[R7.6]   Cerny, M.: Fuzzy Approach to Vector Optimization. Intern. Jour. of General Systems 20(1), 23–29

[R7.7]   Chang, C.L.: Interpretation and Execution of Fuzzy Programs. In: Zadeh, L.A., et al. (eds.) Fuzzy Sets and Their Applications to Cognitive and Decision Processes, pp. 191–218. Academic Press, New York (1975)

[R7.8]   Chang, S.K.: On the Execution of Fuzzy Programs Using Finite State Machines. IEEE Trans. Comp. C-12, 214–253 (1982)

[R7.9]   Chang, S.S.: Fixed Point Theorems for Fuzzy Mappings. Fuzzy Sets and Systems 17, 181–187 (1985)

[R7.10]   Chang, S.S.L.: Fuzzy Dynamic Programming and the Decision Making Process. In: Proc. 3rd Princeton Conference on Information Science and Systems, pp. 200–203. Princeton (1969)

[R7.11]   Chang, S.Y., et al.: Modeling To Generate Alternatives: A Fuzzy Approach. Fuzzy Sets and Systems 9(2), 137–151 (1983)

[R7.12]   Dompere, K.K.: Fuzziness, Rationality, Optimality and Equilibrium in Decision and Economic Theories. In: Lodwick, W.A., Kacprzyk, J. (eds.) Fuzzy Optimization. STUDFUZZ, vol. 254, pp. 3–32. Springer, Heidelberg (2010)

[R7.13]   Dompere, K.K.: On Epistemology and Decision-Choice Rationality. In: Trappl, R. (ed.) Cybernetics and System Research, pp. 219–228. North-Holland, New York (1982)

[R7.14]   Dompere, K.K.: Fuzzy Rationality: Methodological Critique and Unity of Classical, Bounded and Other Rationalities. STUDFUZZ, vol. 235. Springer, New York (2009)

[R7.15]   Dompere, K.K.: Epistemic Foundations of Fuzziness: Unified Theories on Decision-Choice Processes. STUDFUZZ, vol. 236. Springer, New York (2009)

[R7.16]   Dompere, K.K.: Fuzziness and Approximate Reasoning: Epistemics on Uncertainty, Expectations and Risk in Rational Behavior. STUDFUZZ, vol. 237. Springer, New York (2009)

[R7.17]   Dubois, D., et al.: Systems of Linear Fuzzy Constraints. Fuzzy Sets and Systems 3(1), 37–48 (1980)

[R7.18]   Dubois, D.: An Application of Fuzzy Arithmetic to the Optimization of Industrial Machining Processes. Mathematical Modelling 9(6), 461–475 (1987)

[R7.19]   Edwards, W.: The Theory of Decision Making. Psychological Bulletin 51, 380–417 (1954)

[R7.20]   Eaves, B.C.: Computing Kakutani Fixed Points. Journal of Applied Mathematics 21, 236–244 (1971)

[R7.21]   Feng, Y.J.: A Method Using Fuzzy Mathematics to Solve the Vector Maxim Problem. Fuzzy Set and Systems 9(2), 129–136 (1983)

[R7.22]   Hamacher, H., et al.: Sensitivity Analysis in Fuzzy Linear Programming. Fuzzy Sets and Systems 1, 269–281 (1978)

[R7.23]   Hannan, E.L.: On the Efficiency of the Product Operator in Fuzzy Programming with Multiple Objectives. Fuzzy Sets and Systems 2(3), 259–262 (1979)

[R7.24]   Hannan, E.L.: Linear Programming with Multiple Fuzzy Goals. Fuzzy Sets and Systems 6(3), 235–248 (1981)

[R7.25]   Heilpern, S.: Fuzzy Mappings and Fixed Point Theorem. Journal of Mathematical Analysis and Applications 83, 566–569 (1981)

[R7.26]   Ignizio, J.P., et al.: Fuzzy Multicriteria Integer Programming via Fuzzy Generalized Networks. Fuzzy Sets and Systems 10(3), 261–270 (1983)

[R7.27]   Jarvis, R.A.: Optimization Strategies in Adaptive Control: A Selective Survey. IEEE Trans. Syst. Man. Cybernetics SMC-5, 83–94 (1975)

[R7.28]   Jakubowski, R., et al.: Application of Fuzzy Programs to the Design of Machining Technology. Bulleting of the Polish Academy of Science 21, 17–22 (1973)

[R7.29]   Kabbara, G.: New Utilization of Fuzzy Optimization Method. In: Gupta, M.M., et al. (eds.) Approximate Reasoning in Decision Analysis, pp. 239–246. North Holland, New York (1982)

[R7.30]   Kacprzyk, J., et al. (eds.): Optimization Models Using Fuzzy Sets and Possibility Theory. D. Reidel, Boston (1987)

[R7.31]   Kakutani, S.: A Generalization of Brouwer's Fixed Point Theorem. Duke Mathematical Journal 8, 416–427 (1941)

[R7.32]   Kaleva, O.: A Note on Fixed Points for Fuzzy Mappings. Fuzzy Sets and Systems 15, 99–100 (1985)

[R7.33]   Kandel, A.: On Minimization of Fuzzy Functions. IEEE Trans. Comp. C-22, 826–832 (1973)

[R7.34]   Kandel, A.: Comments on Minimization of Fuzzy Functions. IEEE Trans. Comp. C-22, 217 (1973)

[R7.35]   Kandel, A.: On the Minimization of Incompletely Specified Fuzzy Functions. Information, and Control 26, 141–153 (1974)

[R7.36]   Lai, Y., et al.: Fuzzy Mathematical Programming. Springer, New York (1992)

[R7.37]   Leberling, H.: On Finding Compromise Solution in Multcriteria Problems, Using the Fuzzy Min-Operator. Fuzzy Set and Systems 6(2), 105–118 (1981)

[R7.38]   Lee, E.S., et al.: Fuzzy Multiple Objective Programming and Compromise Programming with Pareto Optimum. Fuzzy Sets and Systems 53(3), 275–288 (1993)

[R7.39]   Lowen, R.: Connex Fuzzy Sets. Fuzzy Sets and Systems 3, 291–310 (1980)

[R7.40]   Luhandjula, M.K.: Compensatory Operators in Fuzzy Linear Programming with Multiple Objectives. Fuzzy Sets and Systems 8(3), 245–252 (1982)

[R7.41]   Luhandjula, M.K.: Linear Programming Under Randomness and Fuzziness. Fuzzy Sets and Systems 10(1), 45–54 (1983)

[R7.42]   Negoita, C.V., et al.: Fuzzy Linear Programming and Tolerances in Planning. Econ. Group Cybernetic Studies 1, 3–15 (1976)

[R7.43]   Negoita, C.V., Stefanescu, A.C.: On Fuzzy Optimization. In: Gupta, M.M., et al. (eds.) Approximate Reasoning in Decision Analysis, pp. 247–250. North Holland, New York (1982)

[R7.44]   Negoita, C.V.: The Current Interest in Fuzzy Optimization. Fuzzy Sets and Systems 6(3), 261–270 (1981)

[R7.45]   Negoita, C.V., et al.: On Fuzzy Environment in Optimization Problems. In: Rose, J., et al. (eds.) Modern Trends in Cybernetics and Systems, pp. 13–24. Springer, Berlin (1977)

[R7.46]   Orlovsky, S.A.: On Programming with Fuzzy Constraint Sets. Kybernetes 6, 197–201 (1977)

[R7.47]   Orlovsky, S.A.: On Formulation of General Fuzzy Mathematical Problem. Fuzzy Sets and Systems 3, 311–321 (1980)

[R7.48]   Ostasiewicz, W.: A New Approach to Fuzzy Programming. Fuzzy Sets and Systems 7(2), 139–152 (1982)

[R7.49]   Pollatschek, M.A.: Hieranchical Systems and Fuzzy-Set Theory. Kybernetes 6, 147–151 (1977)

[R7.50]   Ponsard, G.: Partial Spatial Equilibra With Fuzzy Constraints. Journal of Regional Science 22(2), 159–175 (1982)

[R7.51]   Prade, M.: Operations Research with Fuzzy Data. In: Want, P.P., et al. (eds.) Fuzzy Sets, pp. 155–170. Plenum, New York (1980)

[R7.52]   Ralescu, D.: Optimization in a Fuzzy Environment. In: Gupta, M.M., et al. (eds.) Advances in Fuzzy Set Theory and Applications, pp. 77–91. North-Holland, New York (1979)

[R7.53]   Ralescu, D.A.: Orderings, Preferences and Fuzzy Optimization. In: Rose, J. (ed.) Current Topics in Cybernetics and Systems. Springer, Berlin (1978)

[R7.54]   Sakawa, M.: Fuzzy Sets and Interactive Multiobjective Optimization. Plenum Press, New York (1993)

[R7.55]   Sakawa, M., et al.: Feasibility and Pareto Optimality for Multi-objective Nonlinear Programming Problems with Fuzzy Parameters. Fuzzy Sets and Systems 43(1), 1–15

[R7.56]   Tanaka, K., et al.: Fuzzy Programs and Their Execution. In: Zadeh, L.A., et al. (eds.) Fuzzy Sets and Their Applications to Cognitive and Decision Processes, pp. 41–76 (1974)

[R7.57]   Tanaka, K., et al.: Fuzzy Mathematical Programming. Transactions of SICE, 109–115 (1973)

[R7.58]   Tanaka, H., et al.: On Fuzzy-Mathematical Programming. Journal of Cybernetics 3(4), 37–46 (1974)

[R7.59]   Vira, J.: Fuzzy Expectation Values in Multistage Optimization Problems. Fuzzy Sets and Systems 6(2), 161–168 (1981)

[R7.60]   Verdegay, J.L.: Fuzzy Mathematical Programming. In: Gupta, M.M., et al. (eds.) Fuzzy Information and Decision Processes, pp. 231–238. North-Holland, New York (1982)

[R7.61]   Warren, R.H.: Optimality in Fuzzy Topological Polysystems. Jour. Math. Anal. 54, 309–315 (1976)

[R7.62]   Weiss, M.D.: Fixed Points, Separation and Induced Topologies for Fuzzy Sets. Jour. Math. Anal. and Appln. 50, 142–150 (1975)

[R7.63]   Wiedey, G., Zimmermann, H.J.: Media Selection and Fuzzy Linear Programming. Journal Oper. Res. Society 29, 1071–1084 (1978)

[R7.64]   Wilkinson, J.: Archetypes, Language, Dynamic Programming and Fuzzy Sets. In: Wilkinson, J., et al. (eds.) The Dynamic Programming of Human Systems, pp. 44–53. Information Corp., New York (1973)

[R7.65]   Yager, R.R.: Mathematical Programming with Fuzzy Constraints and Preference on the Objective. Kybernetes 8, 285–291 (1979)

[R7.66]   Zadeh, L.A.: Outline of a New Approach to the Analysis of Complex Systems and Decision Process. In: Cochrane, J.L., et al. (eds.) Multiple Criteria Decision Making, pp. 28–44. Univ. of South Carolina Press, Columbia (1973); Also in IEEE Transactions on System, Man and Cybernetics 1, 28–44

[R7.67]   Zadeh, L.A.: The Role of Fuzzy Logic in the Management of Uncertainty in expert Systems. Fuzzy Sets and Systems 11, 199–227 (1983)

[R7.68]   Zimmerman, H.-J.: Description and Optimization of Fuzzy Systems. Intern. Jour. Gen. Syst. 2(4), 209–215 (1975)

[R7.69]   Zimmerman, H.-J.: Fuzzy Programming and Linear Programming with Several Objective Functions. Fuzzy Sets and Systems 1(1), 45–56 (1978)

[R7.70]   Zimmerman, H.J.: Applications of Fuzzy Set Theory to Mathematical Programming. Information Science 36(1), 29–58 (1985)

R8 National Interest and Foreign Policy

[R8.1]   Almond, G.A.: The American people and foreign policy. Harcourt, New York (1950)

[R8.2]   Allen, J.S.: Atomic Imperialism: The State, Monopoly and the Bomb. International Publishers, New York (1952)

[R8.3]   Beard, C.A.: The Idea of National Interest. Macmillan, New York (1934)

[R8.4]   Beloff, M.: The future of British foreign policy. Secker & Warburg, London (1969)

[R8.5]   Boulding, K.E.: The Image. University of Michigan Press, Ann Arbor (1956)

[R8.6]   Burton, J.W.: International Relations; a General Theory. Cambridge
            University Press, London (1965)

[R8.7]   Butterfield, H.: The Scientific v. the Moralistic Approach in International
            Affairs. International Affairs 27, 411–422 (1951)

[R8.8]   Butwell, R. (ed.): Foreign policy and the developing nation: Papers by
            Henry Bienen. University of Kentucky Press, Lexington (1969)

[R8.9]   Cherlesworth, J. (ed.): American Foreign Policy Challenged, Philadelphia,
            The Annals, vol. 342. The American Academy of Political and Social
            Science (1962); Sellin, T., Lambert, R.D. (Gen. eds.)

[R8.10]   Cohen, B.C.: The public's impact on foreign policy. Brown, Boston (1972)

[R8.11]   Cook, T.I., Moos, M.: Power through purpose; the realism of idealism as a
            basis for foreign policy. Johns Hopkins Press, Baltimore (1954)

[R8.12]   Crabb Jr., C.V.: Nonalignment in Foreign Affairs, Philadelphia, The
            Annals, vol. 362. The American Academy of Political and Social Science
            (1965); Sellin, T., Lambert, R.D. (Gen. eds.)

[R8.13]   Dahl, R.A.: Congress and foreign policy. Norton, New York (1964)

[R8.14]   Davenant, C.: Essays Upon the Balance of Power: The Right of Making
            War, Peace and Alliances; and Univeral Monarchy. Knapton, London
            (1701)

[R8.15]   Dennett, R., Turner, R.K. (eds.): Documents on American Foreign
            Relations, vol. XII. Princeton University Press, Princeton (1952)

[R8.16]   Derber, M.: The American idea of industrial democracy, 1865-1965.
            University of Illinois Press, Urbana (1970)

[R8.17]   Destler, I.M.: Making Foreign Economic Policy. The Brookings
            Institutions, Washington, DC (1980)

[R8.18]   Dougherty, J.E., Faltzgraff, P. (eds.): Contending Theories of International
            Relations. Lippincott, Philadelphia (1971)

[R8.19]   Evans, P.B., et al.: Double – Edge Diplomacy. International Bargaining and
            Domestic politics. University of Californian Press, Berkeley (1993)

[R8.20]   Fischman, L.L.: World mineral trends and U.S. supply problems,
            Washington, DC, Resources for the Future. Johns Hopkins University
            Press, Baltimore (1980)

[R8.21]   Galtung, J.: A Structural Theory of Imperialism. Journal of Peace
            Research 2, 81–117 (1971)

[R8.22]   Goldwin, R.A. (ed.): Readings in American Foreign policy. Oxford
            University Press, New York (1971)

[R8.23]   Govett, G.J.S., Govett, M.H. (eds.): World mineral supplies: assessment and
            perspective. Elsevier Scientific Pub. Co., Amsterdam (1976)

[R8.24]   Hanrieder, W.F. (ed.): Comparative foreign policyl theoretical essays.
            McKay, New York (1971)

[R8.25]   Hargreaves, D., Fromson, S.: World index of strategic minerals: production,
            exploitation, and risk. Facts on File, New York (1983)

[R8.26]   Harrod, J.: International Relations: Perceptions and Neo-realism. Year
            Book of World Affairs 31, 289–305 (1977)

[R8.27]   Hartmann, F.H. (ed.): Basic documents of international relations. Kennikat
            Press, Port Washington (1969)

[R8.28]   Hill, M.: A Goals – Achievement Matrix for Evaluating Alternative Plans.
            Journal of Amer. Institute of Planners 34(1), 19–29

[R8.29]   Hinsley, F.H.: Power and the Pursuit of Peace. Cambridge University Press,
            London (1963)

[R8.30]   Hollingsworth, J.R. (ed.): Social Theory and Public Policy, Philadelphia, The Annals, vol. 434. The American Academy of Political and Social Science (1977); Laambert, R.D., Heston, A.W. (Gen. eds.)

[R8.31]   Hoskins, H.L. (ed.): Aiding Underdeveloped Areas Abroad, Philadelphia, The Annals, vol. 268. The American Academy of Political and Social Science (1950); Sellin, T., Charlesworth, J.C. (Gen. eds.)

[R8.32]   Houghton, N.D.: The Challenge to Political Scientists in Recent American Foreign Policy: Scholarship or Ideology? American Political Science Review 52, 678–688 (1958)

[R8.33]   Johnson, R.A.: The administration of United States foreign policy. University of Texas Press, Austin (1971)

[R8.34]   Kaldor, N. (ed.): Conflicts in Policy objective. Oxford University Press, Oxford (1971)

[R8.35]   Kalijarvi, T.V., Merrow, C.E. (eds.): Congress and Foreign Relations, Philadelphia, The Annals, vol. 289. The American Academy of Political and Social Science (1953)

[R8.36]   Kindleberger, C. (ed.): The International Corporation. MIT Press, Cambridge (1970)

[R8.37]   Kelman, H.C.: International Behavior: A Socail-Psychological Analysis. Hol, Rinehart and Winston, New York (1963)

[R8.38]   Kissinger, H.: American foreign policy; three essays. Norton, New York (1969)

[R8.39]   Lauterpacht, H.: The Function of Law in the International Community. Clarendon, Oxford (1933)

[R8.40]   Lieth, C.K., Furness, J.W., Lewis, C.: World Minerals and World Peace. The Brookings Institution, Washington, DC (1943)

[R8.41]   Lowe, C.J.: The reluctant imperialists; British foreign policy, 1878-1902. Macmillan, New York (1969)

[R8.42]   Marshall, C.B.: The exercise of sovereignty: papers on foreign policy. Johns Hopkins Press, Baltimore (1965)

[R8.43]   Merton, R.K.: Social Theory and Social Structure. Free Press, New York (1949)

[R8.44]   Nielsen, W.A.: The great powers and Africa (Published for the Council on Foreign Relations). Praeger Publishers, New York (1969)

[R8.45]   Northedge, F.S.: The International Political System. Faber, London (1976)

[R8.46]   Ornstein, N. (ed.): Changing Congress: The Committee System, Philadelphia, The Annals, vol. 411. The American Academy of Political and Social Science (1974); Sellin, T., Lambert, R.D. (General eds.)

[R8.47]   Palmer, N. (ed.): The National Interest——Alone or with Others? The Annals, vol. 282. The American Academy of Political and Social Science (1952); General Editor Selling, T.

[R8.48]   Paterson, T.G. (ed.): Major Problems in American Foreign Policy: Documents and Essays. Heath and Co., Lexington (1978)

[R8.49]   Patterson, E.M. (ed.): World Government, Philadelphia, The Annals, vol. 264. The American Academy of Political and Social Science (1949); Sellin, T., Lambert, R.D. (eds.)

[R8.50]   Patterson, E.M. (ed.): Formulating a Point Four Program, Philadelphia, The Annals, vol. 270. The American Academy of Political and Social Science (1950); Sellin, T., Charlesworth, J.C. (eds.)

[R8.51]   Dorner, P., El-Shafie, M.A.: Resources and development: natural resource policies and economic development in an interdependent world. University of Wisconsin Press, Croom Helm, Madison, London (1980)

[R8.52]   Petras, J.: Critical Perspectives on Imperialism and Social Class in the Third World. Monthly Review Press, New York (1978)

[R8.53]   Pfaltzgraff, R.L. (ed.): National Security Policy for the 1980s, Philadelphia, The Annals, vol. 457. The American Academy of Political and Social Science (1981); Lambert, R.D., Heston, A.W. (Gen. eds.)

[R8.54]   Prest, A.R.: The Budget and Interpersonal Distribution. Public Finance 23, 80–98 (1968)

[R8.55]   Presthus, R. (ed.): Interest Groups in International Perspective, Philadelphia, The Annals, vol. 413. The American Academy of Political and Social Science (1974); Sellin, T., Lambert, R.D. (Gen. eds.)

[R8.56]   Raushenbush, S. (ed.): The Future of Natural Resources, Philadelphia, The Annals, vol. 281. The American Academy of Political and Social Science (1952); Sellin, T., Lambert, R.D. (Gen. eds.)

[R8.57]   Russett, B.M., Hanson, E.C.: Interest and Ideology; The Foreign Policy Beliefs of American Businessmen. W.H. Freeman, San Francisco (1975)

[R8.58]   Rosenau, J.N. (ed.): International politics and foreign policy; a reader in research and theory. Free Press, New York (1969)

[R8.59]   Rosenau, J.N.: National leadership and foreign policy; a case study in the mobilization of public support. Princeton University Press, Princeton (1963)

[R8.60]   Rourke, F.E.: Bureaucracy and foreign policy. Johns Hopkins University Press, Baltimore (1972)

[R8.61]   Rummel, R.J.: Understanding Conflict and War, vol. I. Sage Publication, New York (1975)

[R8.62]   Schneider, H.: National Objectives and project Appraisal in Developing countries. Development center, OECD, Paris (1975)

[R8.63]   Seelig, L.C.: Resource management in peace and war. National Defense University Press, Washington, DC (1990)

[R8.64]   Sherif, M.: Group Conflict and Cooperation. Routledge and Kegan Paul, London (1966)

[R8.65]   Simon, H.A.: On the concept of organizational goal. Administrative Science Quarterly 9(1), 1–22 (1964)

[R8.66]   Thompson, K.W.: Beyond National Interest: A Critical Evaluation of Reinhold Neibuh's Theory of International Politics. Review of Politics 13, 379–391 (1956)

[R8.67]   Van Rensburg, W.C.J.: Strategic minerals. Prentice-Hall, Englewood Cliffs (1986)

[R8.68]   Leontief, W., et al.: The Future of nonfuel minerals in the U.S. and world economy: input-output projections, 1980-2030. Lexington Books, Lexington (1983)

[R8.69]   Westerfield, H.B.: Foreign policy and party politics: Pearl Harbor to Korea. Octagon Books, New York (1972)

[R8.70]   Wilson, J.H.: American business & foreign policy. Beacon Press, Boston (1973)

[R8.71]   Young, R.G.: Goals and Goal-Setting. Jour of Amer. Institute of Planners, 76–85 (March 1966)

R9   Negotiations and Resolutions in Conflicts

[R9.1]   Deutsch, M.: The Resolution of conflict: constructive and Destructive process. Conn. Yale University Press, New Haven (1977)

[R9.2]   Dixit, A., Nalebeff, B.: Thinking Strategically. Norton, New York (1991)

[R9.3]   Raiffa, H., et al.: The Science and Art of collaborative Decision. The
            Belknap – Harvard University Press, Cambridge (2002)
[R9.4]   Raiffa, H., et al.: Negotiation Analysis: The Science and Art of
            collaborative
[R9.5]   Mnookin, R.H.: Beyond Winning. Negotiate to create value in Deals and
            Disputes. Harnerd Press, Cambridge (2000)
[R9.6]   Mnookin, R.H.: Beyond Winning. Negotiate to create value. In: Deals and
            Disputes, pp. 203–211. Harnerd Press, Cambridge (1979, 2000)
[R9.7]   Zartman, W.I. (ed.): The Negotiation process: Theories and applications.
            Calif. Sage Pub., Beverly Hill (1978)
[R9.8]   Zeckhouser, R.J., et al. (eds.): Wise Choices: Decisions. Harvard Business
            School Pub., Harvard University Press, Cambridge Mass (1996)

R10 Political Decisions, Games and Rationality

[R10.1]   Adelman, I., Cynthia, T.M.: Society, Politics and Economic Development.
            John Hopkins Press, Baltimore (1967)
[R10.2]   Austen-Smith, D.: Sophisticated Sincerity: Voting over Endogenous
            Agendas. American Political Science Review 81, 1323–1330 (1987)
[R10.3]   Banks, J.S.: Sophisticated Voting Outcomes and Agenda Control. Social
            Choice and Welfare 1, 295–306 (1985)
[R10.4]   Baron, D.P.: Electoral Competition with Informed and Uninformed Voters.
            American Political Science Review 88, 1–14 (1994)
[R10.5]   Baron, D.P.: A Dynamic Theory of Collective Goods Provision. American
            Political Science Review 90, 216–330 (1996)
[R10.6]   Baron, D.P., Ferejohn, J.: Bargaining in Legislatures. American Political
            Science Review 83, 1181–1206 (1989)
[R10.7]   Bernard, J.: The Theory of Games of Strategy as a Modern Sociology of
            Conflict. American Journal of Sociology 59, 411–424
[R10.8]   Bernheim, B.D., Rangel, A., Rayo, L.: The Power of the Last Word in
            Legislative Policy Making. Econometrica 74(5), 1161–1190 (2006)
[R10.9]   Brams, S.J.: Game Theory and Politics. Free Press, New York (1975)
[R10.10]   Brams, S.J.: The Presidential Election Game. Yale University Press, New
            Haven (1978)
[R10.11]   Cowart, A.T., Brofass, K.E.: Decision, Politics, and Change.
            Universitetsforlaget, Oslo (1979)
[R10.12]   Dompere, K.K.: The Social Goal-objective Formation and National Interest
            in Political Economies Under Fuzzy Rationality, A Working Monograph.
            Department of Economics, Howard University, Washington DC (2009)
[R10.13]   Dompere, K.K.: Fuzziness, and the Market Mockery of Democracy: The
            Political Economy of Rent-Seeking and Profit-Harvesting, A Working
            Monograph. Department of Economics, Howard University, Washington
            DC (2010)
[R10.14]   Dompere, K.K.: African Union: Pan-African Analytical Foundations.
            Adonis& Abbey Publishers, London (2006)
[R10.15]   Dompere, K.K.: Polyrhythmicity: Foundations of African Philosophy.
            Adonis& Abbey Publishers, London (2006)
[R10.16]   Epple, D., Riordan, M.H.: Cooperation and Punishment Under Repeated
            Majority Voting. Public Choice 55, 41–73 (1987)

[R10.17]   Mc Kelvey, R.: Intransitivities in Multidimensional Voting Models and
              Some Implication of Agenda Control. Journal of Economic Theory 12, 472–
              482 (1976)
[R10.18]   Noll, R.G., Owen, B.M.: The Political Economy of Deregulation: Interest
              Groups in the Regulatory Process. American Enterprise Institute for Public
              Policy Research, Washington, DC (1983)
[R10.19]   Rapoport, A.: Fights, Games and Debates. Michigan University Press,
              Lansing (1963)
[R10.20]   Romer, R., Rosenthal, H.: Political Resource Allocation, Controlled
              Agendas and the Status quo. Public Choice 33, 27–43

R11   Public-Private Sector Relative Size

[R11.1]   Arrow, K.J., Boskin, M.J. (eds.): The Economics of Public Debt. Macmillan
              Press (1988)
[R11.2]   Aschauer, D.A.: Is Public Expenditure Productive? Journal of Monetary
              Economics 23, 177–200 (1989)
[R11.3]   Baumol, W.J. (ed.): Public and Private Enterprise in a Mixed Economy
              (Proceedings of Conference held by the International Economic Association
              in Mexico City). St. Martin's Press, New York (1980)
[R11.4]   Beck, M.: Public Sector Growth: A Real Perspective. Public Finance 34,
              10–27 (1979)
[R11.5]   Baird, C.W.: On Profits and Hospitals. Journal of Economic Issues 5, 57–66
              (1971)
[R11.6]   Balkan, E., Greene, K.V.: On Democracy and Debt. Public Choice 67, 201–
              211 (1990)
[R11.7]   Baumol, W.J. (ed.): Public and Private Enterprise in a Mixed Economy
              (Proceedings of Conference held by the International Economic Association
              in Mexico City). St. Martin's Press, New York (1980)
[R11.8]   Barro, R.: On the Determination of the Public Debt. Journal of Political
              Economy 87, 940–971 (1979)
[R11.9]   Borcherding, T.E.: Budgets and Bureaucrats: The Sources of Government
              Growth. Duke University Press, Durham (1977)
[R11.10]   Buchanan, J.M., Rowley, C.K., Tollison, R.D. (eds.): Deficits. Basil
              Blackwell, Oxford (1987)
[R11.11]   Butler, S.M.: Privatizing Federal Spending: A Strategy to Eliminate the
              Deficit. Universal Books, New York (1985)
[R11.12]   Cameron, D.R.: The Expansion of the Public Economy: A Comparative
              Analysis. American Political Science Review 72, 1243–1261 (1978)
[R11.13]   Clarke, S.R.: The Management of the Public Sector of the National
              Economy. Athlone Press, London (1964)
[R11.14]   Coombes, D.: State Enterprise: Business or Politics? Allen and Unwin,
              London (1970)
[R11.15]   Davies, D.G.: The Efficiency of Public Versus Private Firms, The Case of
              Australia's Two Airlines. Journal of Law and Economics 14(31), 149–165
              (1971)
[R11.16]   Eisner, R.: How real Is the Federal Deficits. The Free Press, A division of
              Macmillan, New York
[R11.17]   Eisner, R.: Budget Deficits: Rhetoric and Reality. Journal of Economic
              Perspectives 3, 72–93 (1989)

[R11.18]   Eisner, R., Pieper, P.J.: Deficits, Monetary Policy and Real Economic Activity. In: Arrow, K.J., Boskin, M.J. (eds.) The Economics of Public Debt, pp. 3–40. Macmillan Press, New York (1988)

[R11.19]   Eisner, R., Pieper, P.J.: A New View of the Federal Debt and Budget Deficits. American Economic Review 74, 11–29 (1984)

[R11.20]   Feldman, A.M.: A Model of majority Voting and Growth in Government Expenditure. Public Choice 46(i), 3–17 (1985)

[R11.21]   Fitch, L.C.: Increasing the Role of the Private Sector in Providing Public Services. In: Hawley, W.D., Rogers, D. (eds.) Improving the Quality of Urban Management, pp. 501–559. Sage, Beverly Hills (1974)

[R11.22]   Gramlich, E.M.: U. S. Federal Budget Deficits and Gramm-Rudman-Holdings. American Economic Review 80, 75–80 (1990)

[R11.23]   Galbraith, J.K.: Economics and Public Purpose. Houghton Mifflin, Boston (1973)

[R11.24]   Hanke, S.H.: Prospects for Privatization. Academy of Political Science, New York (1987)

[R11.25]   Hanson, A. (ed.): Organization and Administration of Public Enterprise. United Nations, New York (1968)

[R11.26]   Hayek, F.A.: New Studies in Philosophy. Politics, Economics and History of Ideas. University of Chicago Press, Chicago (1978)

[R11.27]   Heilbroner, R.L., Bernstein, P.L.: The Debt and Deficit; False Alarms/Real Possibilities. Norton, New York (1990)

[R11.28]   Heclo, H., Wildavsky, A.: The Private Government of Public Money: Community and Policy Inside British Politics. Macmillan, London (1974)

[R11.29]   Hsiao, W.: Public Versus Private Administration of Health Insurance: A Study in Relative Economic Efficiency. Inquiry 15, 379–387 (1978)

[R11.30]   Johnson, R.T.: Historical Beginnings ...The Federal Reserve. Federal Reserve Bank of Boston, Boston (1977)

[R11.31]   Kaldor, N.: Public or Private Enterprise- The Issues to be Considered. In: Baumol, W.J. (ed.) Public and Private Enterprise in a Mixed Economy (Proceedings of Conference held by the International Economic Association in Mexico City), pp. 1–12. St. Martin's Press, New York (1980)

[R11.32]   Kendrick, J.W.: The Formation and Stocks of Total Capital. Columbia University Press, New York (1976)

[R11.33]   Kimmel, L.H.: Federal Budget and Fiscal policy, 1789-1958. Brookings Institution, Washington DC (1959)

[R11.34]   Kotlikoff, L.J.: Deficit Delusion. Public Interest 84, 53–65 (1984)

[R11.35]   Kuttner, R.: The Economic Illusion: False Choices Between Prosperity and Social Justices. Houghton Mifflin, Boston (1984)

[R11.36]   Kotlikoff, L.J.: The Deficit Is Not a Well-Defined Measure of Fiscal Policy. Science 241, 791–795 (1988)

[R11.37]   Lindsay, C.M.: A Theory of Government Enterprise. Journal of Political Economy 84, 1061–1077 (1976)

[R11.38]   Meltzer, A.H., Richard, S.F.: Why Government Grows (and Grows) in Democracy. Public Interest 52, 111–118 (1978)

[R11.39]   Meyer, R.A.: Publically Owned Versus Privately Owned Utilities: A Policy Choice. Review of Economics and Statistics 57(4), 391–399 (1975)

[R11.40]   Naylor, R.T.: Hot Money and the Politics of Debt. Simon and Schuster, New York (1987)

[R11.41]   Peacock, A.T., Wiseman, J.: The Growth of Public Expenditure in the United Kingdom, 1890-1955. Princeton University Press, Princeton (1961)

[R11.42]   Peacock, A., Shaw, G.K.: The economic Theory of Fiscal Policy. St Martin's Press, New York (1976)

[R11.43]   Peston, M.: Public Goods and the Private Sector. Macmillan, London (1972)

[R11.44]   Phelps, E.S. (ed.): Private Wants and Public Needs. W. W. Norton, New York (1957)

[R11.45]   Reich, R.: The Power of Public Ideas. Ballinger Pub., Cambridge (1987)

[R11.46]   Rock, J.M. (ed.): Debt and the Twin Deficits Debate. Mayfield Pub., Mountain View (1991)

[R11.47]   Runge, C.F.: The Fallacy of Privatization. Journal of Contemporary Studies, 3–17 (Winter 1984)

[R11.48]   Saunders, P.: Public Expenditure and Economic Performance in OECD Countries. Journal of Public Policy 5, 1–21 (1986)

[R11.49]   Savas, E.S.: Public vs. Private Refuse Collection: A Critical Review of the Evidence. Journal of Urban Analysis 6, 1–13 (1979)

[R11.50]   Savas, E.S.: Privatizing the Public Sector: How to Shrink Government. Chatham House Publishers, Inc., Chatham (1982)

[R11.51]   Schick, A.: Congress and Money: Budgeting, Spending and Taxing. The Urban Institute, Washington, DC (1980)

[R11.52]   Shonfield, A.: Modern Capitalism: The Changing Balance of Public and Private Power. Oxford University Press, New York (1965)

[R11.53]   Spann, R.M.: Public Versus Private Provision of Government Services. In: Borcherding, T.E. (ed.) Budgets and Bureaucrats: The Sources of Government Growth, pp. 82–87. Duke University Press, Durham (1977)

[R11.54]   Wagner, R.E.: Revenue Structure, Fiscal Illusion, and Budgetary Choice. Public Choice 25, 45–61 (1976)

[R11.55]   Watson, D.S.: Economic Policy: Business and Government. Houghton Mifflin, Boston (1960)

[R11.56]   Wilson, J.Q., Rachal, P.: Can the Government Regulate Itself? Public Interest 46, 3–14 (1977)

[R11.57]   Yunker, J.E.: Economic Performance of Public and Private Enterprise: The Case of U. S. Electric Utilities. Journal of Economics and Business 28, 60–67 (1975)

R12   Regulation and Deregulation Game

[R12.1]   Beesley, M.E.: Privatization, Regulation and Deregulation. Routledge, London (1992)

[R12.2]   Buchanan, J.M.: A Contractual Paradigm for Applying Economic Theory. American Economic Review 65, 225–230 (1975)

[R12.3]   Cavaco-Silva, A.: Economic Effects of Public Debt. St. Martin's Press, New York (1977)

[R12.4]   Gayle, D.J., Goodrich, J.N. (eds.): Privatization and deregulation in global perspective. Quorum Books, New York (1990)

[R12.5]   Fowler, R.B.: Energy & the Deregulated Marketplace: 1998 survey. The Fairmont Press, Lilburn (1998)

[R12.6]   Landy, M.K., Levine, M.A., Shapiro, M. (eds.): Creating Competitive Markets: The Politics of Regulatory Reform. Brookings Institution Press, Washington, DC (2007)

[R12.7]   Letwin, W.: Law and Economic Policy in America: The Evolution of Sherman Antitrust Act. Random House, New York (1965)

[R12.8]   Milanovic, B.: Liberalization and Entrepreneurship: Dynamics of Reform in Socialism and Capitalism. M. E. Sharp, Inc., New York (1989)
[R12.9]   Peltzman, S.: Toward a More General Theory of Regulation. Journal of Law and Economics, 211–240 (1976)
[R12.10]   Phillips Jr., C.F.: The Economics of Regulation. Irwin, Holmwood (1969)
[R12.11]   Posner, R.A.: The Social Cost of Monopoly and Regulation. Journal of Political Economy 83, 507–827 (1975)
[R12.12]   Spulber, D.F.: Regulation and Merkets. MIT Press, Cambridge (1989)
[R12.13]   Swary, I., Topf, B.: Global Financial Deregulation: Commercial Banking at the Crossroads. USA Blackwell, Cambridge (1992)

R13   Rent-Seeking and Pork Barreling

[R13.1]   Buchanan, J.M., et al. (eds.): Toward a theory of the Rent seeking society. Taxas A and M University Press, college station (1980)
[R13.2]   Ferejohn, J.: Pork Barrel Politics: Rivers and Harbors Legislation, 1947-1968. Stanford University Press, Stanford (1974)
[R13.3]   Krueger, A.: The Political Economy of Rent seeking society. American Economic Review 64, 291–302 (1974)
[R13.4]   Nti, K.O.: Rent-Seeking with Asymmetric Valuation. Public Choice 98, 415–430 (1999)
[R13.5]   Tollison, R.D.: Rent-Seeking: A Survey. Kyklos 35, 575–602 (1982)
[R13.6]   Tullock, G.: The Rent-Seeking Society: The Selected Works, vol. 5. Liberty Fund, Indianapolis (2005)

R14   Government Revenues, Taxes and Expenditures

[R14.1]   Arrow, K.J.: Equality in Public Expenditure. Quarterly Jour. of Econ. 85, 409–415 (1971)
[R14.2]   Dorfman, R. (ed.): Measuring Benefits of Government Investments. The Brookings Institution, Washington, DC (1965)
[R14.3]   Downs, G.W., Larkey, P.D.: The Search for Government Efficiency. Temple University Press, Philadelphia (1986)
[R14.4]   Else, P.K., Marshall, G.P.: The Management of Public Expenditure. Policy Studies Institute, London (1979)
[R14.5]   Gerwin, D.: Towards a Theory of Public Budgetary Decision Making. In: Byrne, R.F., et al. (eds.) Studies in Budgeting. North-Holland, Amsterdam (1971)
[R14.6]   Oates, W.E. (ed.): The Political Economy of Fiscal Federalism. Lexington Press, Lexington (1977)
[R14.7]   Peacock, A.: The Economic Analysis of Government and Related Themes. St. Martins's Press, New York (1979)
[R14.8]   Schultze, C.L.: The Politics and Economics of Public Spending. The Brookings Institution, Washington, DC (1968)

R15   Some Relevant Revolutions in Thought and Political Economy

[R15.1]   Arrow, K.J.: Limited Knowledge and Economic Analysis. American Economic Review 64, 1–10 (1974)

[R15.2]   Arrow, K.J.: General Economic Equilibrium: Purpose, Analytic Techniques, Collective Choice. American Economic Review 64, 253–272 (1974)

[R15.3]   Black, M.: The Nature of Mathematics. Littlefield, Adams and Co., Totowa (1965)

[R15.4]   Blass, A.: The Interaction Between Category and Set Theory. Mathematical Applications of Category Theory 30, 5–29 (1984)

[R15.5]   Brouwer, L.E.J.: Intuitionism and Formalism. Bull of American Math. Soc. 20, 81–96 (1913); Also in Benecerraf, P., Putnam, H. (eds.): Philosophy of Mathematics: Selected Readings, pp. 77–89. Cambridge University Press, Cambridge (1983)

[R15.6]   Brouwer, L.E.J.: Consciousness, Philosophy, and Mathematics. In: Benecerraf, P., Putnam, H. (eds.) Philosophy of Mathematics: Selected Readings, pp. 90–96. Cambridge University Press, Cambridge (1983)

[R15.7]   Brown, B., Woods, J. (eds.): Logical Consequence; Rival Approaches and New Studies in exact Philosophy: Logic, Mathematics and Science, vol. II. Hermes, Oxford (2000)

[R15.8]   Campbell, N.R.: What is Science? Dover, New York (1952)

[R15.9]   Carnap, R.: Foundations of Logic and Mathematics. In: International Encyclopedia of Unified Science, pp. 143–211. Univ. of Chicago, Chicago (1939)

[R15.10]   Carnap, R.: The Two Concepts of Probability. Philosophy and Phenomenonological Review 5, 513–5532 (1945)

[R15.11]   Carnap, R.: The Methodological Character of Theoretical Concepts. In: Feigl, H., Scriven, M. (eds.) Minnesota Studies in the Philosophy of Science, vol. I, pp. 38–76 (1956)

[R15.12]   Carson, R.B., Ingles, J., McLand, D. (eds.): Government in the American Economy: Conventional and Radical Studies on the Growth of State Economic Power. Lexington Press, Lexington (1977)

[R15.13]   Dompere, K.K.: The Theory of the Knowledge Square: The Analytical Foundations of Knowing, A Working Monograph on Philosophy of Science I. Howard University, Washington, DC (2011)

[R15.14]   Dompere, K.K.: Fuzzy Rational Foundations of Exact and Inexact Sciences, A Working Monograph on Philosophy of Science II. Howard University, Washington, DC (2011)

[R15.15]   Dompere, K.K.: Fuzziness, Rationality, Optimality and Equilibrium in Decision and Economic Theories. In: Lodwick, W.A., Kacprzyk, J. (eds.) Fuzzy Optimization. STUDFUZZ, vol. 254, pp. 3–32. Springer, Heidelberg (2010)

[R15.16]   Dompere, K.K.: On Epistemology and Decision-Choice Rationality. In: Trappl, R. (ed.) Cybernetics and System Research, pp. 219–228. North-Holland, New York (1982)

[R15.17]   Dompere, K.K.: Fuzziness and the Market Mockery of Democracy: The Political Economy of Rent-Seeking and Profit-Harvesting, A Working Monograph, Department of Economics. Howard University, Washington DC (2009)

[R15.18]   Dompere, K.K.: Social Goal-Objective Formation, Democracy and National Interest: Political Economy Under Fuzzy Rationality, A Working Monograph, Department of Economics. Howard University, Washington DC (2009)

[R15.19]   Dretske, F.I.: Knowledge and the Flow of Information. MIT Press, Cambridge (1981)

[R15.20]   Harwood, E.C.: Reconstruction of Economics. Amarican Institute of Economic Research, Great Barrington (1955)

[R15.21]   Hayek, F.A.: New Studies in Philosophy, Politics, Economics and the History of Ideas. The University of Chicago Press, Chicago (1978)

[R15.22]   Helmer, O., Resher, N.: On the Epistemology of Inexact Science, P-1513. Rand Corporation (October 13, 1958)

[R15.23]   Kay, G.: The Economic Theory of the Working Class. St. Martin's Press, New York (1979)

[R15.24]   Keirstead, B.S.: The Conditions of Survival. American Economic Review 40(2), 435–445

[R15.25]   Knight, F.H.: On the History and Method of Economics. University of Chicago Press, Chicago (1966)

[R15.26]   Kuznets, S.: Toward a Theory of Economic Growth. Norton, New York (1968)

[R15.27]   Kühne, K.: Economics and Marxism. The Renaissance of the Marxian System, vol. I. St. Martin's Press, New York (1979)

[R15.28]   Kühne, K.: Economics and Marxism. The Dynamics of the Marxian System, vol. II. St. Martin's Press, New York (1979)

[R15.29]   March, J.G.: Bounded Rationality, Ambiguity and Engineering of Choice. The Bell Journal of Economics 9(2), 587–608 (1978)

[R15.30]   Marx, K.: The Poverty of Philosophy. International Publishers, New York (1963)

[R15.31]   Mátyás, A.: History of Modern Non-Marxian Economics: From Marginalist Revolution through Keynesian Revolution to Contemporary Monetarist Counter- revolution. Akademiai Kiodo, Budapest (1980)

[R15.32]   Niebyl, K.H.: Modern Mathematics and Some Problems of Quality, Quantity and Motion in Economic Analysis. Philosophy of Science 7(31), 103–120 (1940)

[R15.33]   Nkrumah, K.: Consciencism. Modern Reader, New York (1970)

[R15.34]   Pollock, J.: Knowledge and Justification. Princeton University Press, Princeton (1974)

[R15.35]   Polanyi, M.: Personal Knowledge. Routledge and Kegan Paul, London (1958)

[R15.36]   Price, H.H.: Thinking and Experience. Hutchinson, London (1953)

[R15.37]   Popper, K.R.: Objective Knowledge. Macmillan, London (1949)

[R15.38]   Putman, H.: Reason, Truth and History. Cambridge University Press, Cambridge (1981)

[R15.39]   Putman, H.: Realism and Reason. Cambridge University Press, Cambridge (1983)

[R15.40]   Quiggin, J.: Zombie Economics: How Dead Ideas Still Walk among Us. Princeton University Press, Princeton (2010)

[R15.41]   Robinson, J.: Economic Philosophy. Anchor Books, New York (1962)

[R15.42]   Robinson, J.: Freedom and Necessity: An Introduction to the Study of Society. Vintage Books, New York (1971)

[R15.43]   Robinson, J.: Economic Heresies: Some Old-Fashioned Questions in Economic Theory. Basic Books, New York (1973)

[R15.44]   Simon, H.A.: Models of Bounded Rationality. MIT Press, Cambridge (1982)

[R15.45]   Tigar, M.E., Levy, M.R.: Law and the Rise of Capitalism. Monthly Review Press, New York (1977)

Subject Index

Printed in the United States
By Bookmasters